Contemporary Genetics
Laboratory Manual

Rodney J. Scott
Wheaton College

ISBN: 978-1-60250-086-0
         20110003

TABLE OF CONTENTS

LAB ACTIVITY 1

Caenorhabditis elegans
A Tiny Worm That's a Major Model System

INTRODUCTION

In December 1998, researchers announced a milestone in genome analysis. The sequencing of the 97 million base pairs of *Caenorhabditis elegans* was essentially completed. With this achievement, the nematode *C. elegans* became the first multicellular organism for which a complete genome sequence was available.

Since the 1970s, *C. elegans* ("the worm," as it's known to those who study it) has been a favorite model organism of developmental biologists and geneticists. A key characteristic that makes it so attractive for studies of development is the small number of cells that comprise its body — an adult hermaphrodite has exactly 959 somatic cells. Since the worm is transparent, its development can be carefully followed from a single cell to a mature adult. Microscopic studies of this process have traced the exact cell lineage of each of the 959 somatic cells.

Furthermore, mutant *C. elegans* can be isolated with relative ease, and thousands of mutant strains are available for study. This has enabled geneticists to "genetically dissect" the development and physiology of the worm. As a result, *C. elegans* serves as an important model organism for studies involving many complex phenomena associated with multicellular organisms. For example, much of what we now know about the important process of programmed cell death (apoptosis) comes from studies of mutant strains of *C. elegans*.

CHARACTERISTICS OF *C. ELEGANS*

Caenorhabditis elegans is a small soil nematode that eats bacteria and normally reproduces as a hermaphrodite. Morphologically distinct males occur infrequently (< 1% of worms in cultures of wild-type strains are males), but they are unnecessary for reproduction. Hermaphrodites have two X chromosomes, and the rare males are produced when a nondisjunction event produces an embryo with only one X.

Hermaphrodites produce offspring by self-fertilization, or a hermaphrodite may be fertilized by a male. It takes only 2 to 3 days for a fertilized egg to develop into an adult. During that time, four larval stages occur, with a molt between each stage. The morphology of the adult is fairly simple (see Figure 1-1). The body of the worm is surrounded by an acellular cuticle. Within the cuticle, the body consists of gonads, an intestine, muscles, nerves, and hypodermal cells that produce the cuticle.

During this laboratory period, you will become familiar with the normal characteristics of typical male and hermaphrodite worms. You will also be introduced to several very strange mutant strains.

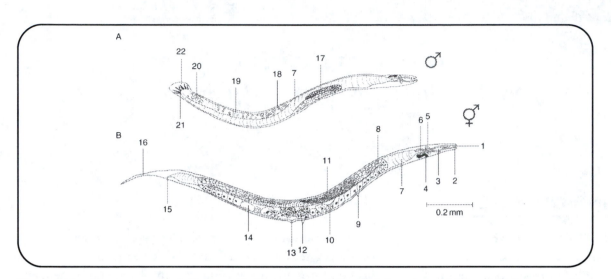

Figure 1.1 – The anatomy of *Caenorhabditis elegans:* (a) an adult male and (b) an adult hermaphrodite. 1. Buccal cavity. 2. One of six lips. 3. Metacropus of the pharynx. 4. Excretory cell. 5. Nerve ring. 6. Terminal bulb of the pharynx. 7. Intestine. 8. Distal arm of the gonad, ovary. 9. Loop of the gonad. 10. Proximal arm of the gonad with maturing oocytes. 11. Spermatheca. 12. Uterus with cleaving eggs. 13. Vulva. 14. Mature oocyte before spermatheca. 15. Anus. 16. Tail. 17. Testis. 18. Spermatocytes. 19. Vas deferens. 20. Mature sperm. 21. Cloaca. 22. Copulatory bursa with rays and copulatory spicules.

PROCEDURE

Microscopic Observations — Living Worms

Obtain a culture of worms labeled "him" (high incidence of males), and view these under the dissecting microscope. This strain of *C. elegans* is morphologically wild-type; however, its cultures yield an abnormally high percentage of males. Observe these living worms and answer the following questions:

 Q1 **Can you distinguish the anterior versus the posterior end of the worm? How? (Describe what you see.)**

Q2 **How are males morphologically different from hermaphrodites? (Do not discuss the internal anatomy in your answer to this question.)**

Q3   How do males and hermaphrodites interact? (Describe what you see.)

Microscopic Observations — Stained Worms

Prepare a slide with several hermaphrodite and several male worms. Place a drop of dilute methylene blue stain on a glass slide, and transfer the worms into the dilute stain. You will have to work carefully, using the dissecting microscope to pick up the worms (using a probe made with a piece of fine wire). When you place the worms in the stain, they will remain alive for some time. However, within 10 to 15 minutes, they will die or become extremely sluggish so that you can make detailed observations using a compound microscope.

As you work with the slide, make sure that there is enough stain present to keep the coverslip floating above the worms. You may need to add more stain from time to time to prevent the coverslip from crushing the worms.

Draw a sketch of a hermaphroditic worm and another one of a male worm. Make your sketches at least 10 to 15 cm in diameter, and include as much detail as possible. On the appropriate sketch(es), label each of the following items: mouth, vulva, copulatory bursa, ovary, and uterus. You may see additional recognizable features, as well.

Observing Mutant Strains

Several mutant strains are available in separate Petri plates. These plates are labeled with code designations, but not the names of the mutant phenotypes. You should carefully observe these worms using a dissecting microscope and determine which phenotypes are represented by each labeled culture. The following are descriptions of some of the possible phenotypes that may be present in the strains available in your lab. Your instructor may provide names of other mutant phenotypes and descriptions besides those listed here.

dumpy — Dumpy mutants are shorter than the wild-type animals, and, as a result, they certainly do look "dumpy."

him (high incidence of males) — As described earlier, strains with this phenotype look normal, but have many more males than do wild-type strains.

long — These are longer and thinner than the wild-type worms.

roller — Instead of moving forward with an "elegant," sinuous movement, these worms roll sideways, moving by rotating around the long axes of their bodies.

shaker — These worms are somewhat paralyzed.

uncoordinated — There are various kinds of "uncoordinated" mutants. This particular strain is known for not being able to "back up." To see this phenotype, try poking a worm near its head. A normal worm will glide elegantly backwards; an uncoordinated worm will just coil up.

wild-type — One of the coded plates will contain normal worms. These will look like the worms illustrated in Figure 1.1 and the living him worms that you sketched earlier. However, these will have an extremely low frequency of male worms (<1%).

Q4 List the code designations of each culture that you observed, and indicate the name of the mutant phenotype exhibited in each case.

NOTES

Ceratopteris richardii
A Versatile Plant Model System

INTRODUCTION

The fern *Ceratopteris richardii* is a simple plant with many features in common with more complex plants. It has a relatively short life cycle and can be easily grown in the laboratory in huge numbers, under highly controlled conditions. Moreover, it exhibits many fascinating characteristics associated with physiological, genetic, developmental, and environmental phenomena. In short, *C. richardii* is a very useful model plant system.

To appreciate fully what makes *Ceratopteris* such an ideal model organism, you must begin by understanding some basics of its life cycle (see Figure 2.1). *Ceratopteris* starts life as a spore. Spores are tiny, single cells that remain "alive" but metabolically inert for years or even decades. *Ceratopteris* researchers typically have vials full of millions of these tiny "preorganisms" lining the shelves of their laboratories. This is a very convenient way to keep a model organism in storage! The gametophytes grow quickly and are extremely small, both characteristics that make it useful for experimental work. Furthermore, both the gametophytic and sporophytic stages of the life cycle have unique attributes that make *Ceratopteris* useful for genetic research (which we will discuss more fully later).

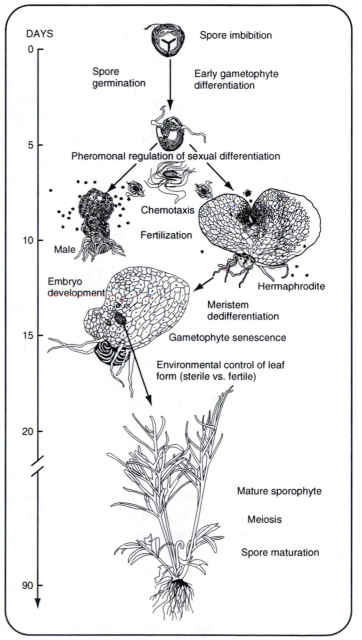

Figure 2.1 – The *Ceratopteris* life cycle. See text for description.

THE GAMETOPHYTE GENERATION

The spore is the first cell of the gametophyte generation (see Figure 2.2 for details of spore and gametophyte morphology). It begins germinating when it is moistened and exposed to light. The single, haploid cell divides by mitosis, and once enough growth has occurred, the cells break through the spore wall. A mass of green tissue that protrudes from the spore is the beginning of the body of the gametophyte. Long, thin, tubular extensions of some of these cells form the rhizoids of the gametophyte.

The gametophytic plants represent one of the two independent stages of the *Ceratopteris* life cycle (the other stage is the sporophyte, discussed later). Mature gametophytes of *Ceratopteris* can be of two types, hermaphroditic or male (see Figure 2.2). Although any spore has the potential to develop as a hermaphrodite, some become males when they grow in the presence of a pheromone called *antheridiogen* (produced by the gametophytes that developed before them). The hermaphrodites are larger than the males and have a mitten-shaped appearance. They produce two types of sex organs: *archegonia* (located in the "notch" region of the gametophyte), which contain eggs, and small, spherical *antheridia*, which contain sperm. The antheridia of the hermaphrodites are scattered mostly on the margins of the gametophytes. The males are smaller, have a club-shaped appearance, and are covered with antheridia. When gametophytes become sexually mature, they are less that 1 mm wide. Hundreds of these tiny plants can easily be grown in a single, small, Petri plate. This makes *Ceratopteris* a very convenient model organism. An experiment that could be done with *Ceratopteris* on a single shelf of a culture chamber might require several acres if another model plant such as corn or tobacco were used!

THE *CERATOPTERIS* LIFE CYCLE

When gametophytes are exposed to water, mature antheridia will burst and release swarms of swimming sperm. The sperm are attracted to the archegonia via a chemical signal that allows them to locate and fertilize the eggs. The diploid cell resulting from the union of a sperm and an egg is the first cell of the sporophyte generation. Initially, the sporophyte is very small and grows out of the gametophyte (see detail in Figure 2.2). Eventually, it becomes a large plant (about the size of a begonia), capable of producing millions of spores every few months.

CERATOPTERIS GENETICS

Because of its unique life cycle, *Ceratopteris* is an especially useful model organism for genetic studies. The tiny haploid spores are ideal for inducing and selecting mutations. Because spores are so small, hundreds of thousands of them can be mutagenized and screened at one time. Because they are haploid, even recessive mutations can be identified. This results in a highly powerful mutant selection system. In a mutant selection scheme, mutagenized gametophytes are grown under conditions where a non-wild-type phenotype can be easily identified (e.g., a "large," healthy, green gametophyte growing among thousands of dead ones on medium supplemented with a herbicide would be one that exhibits a non-wild-type phenotype). Such gametophytes can be isolated and used to generate new mutant strains (described later) if the phenotype is truly the result of a mutation.

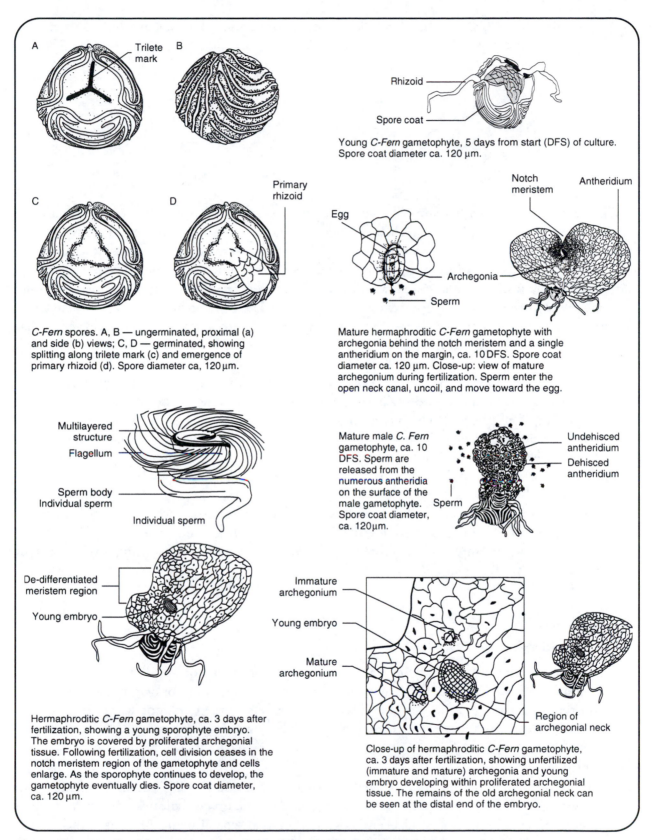

A Trilete mark B

Rhizoid

Spore coat

Young *C-Fern* gametophyte, 5 days from start (DFS) of culture. Spore coat diameter ca. 120 μm.

C D Primary rhizoid

Notch meristem Antheridium

Egg

Archegonia

Sperm

C-Fern spores. A, B — ungerminated, proximal (a) and side (b) views; C, D — germinated, showing splitting along trilete mark (c) and emergence of primary rhizoid (d). Spore diameter ca, 120 μm.

Mature hermaphroditic *C-Fern* gametophyte with archegonia behind the notch meristem and a single antheridium on the margin, ca. 10 DFS. Spore coat diameter ca. 120 μm. Close-up: view of mature archegonium during fertilization. Sperm enter the open neck canal, uncoil, and move toward the egg.

Multilayered structure

Flagellum

Sperm body
Individual sperm

Individual sperm

Mature male *C. Fern* gametophyte, ca. 10 DFS. Sperm are released from the numerous antheridia on the surface of the male gametophyte. Spore coat diameter, ca. 120 μm.

Undehisced antheridium

Dehisced antheridium

Sperm

De-differentiated meristem region

Young embryo

Immature archegonium

Young embryo

Mature archegonium

Region of archegonial neck

Hermaphroditic *C-Fern* gametophyte, ca. 3 days after fertilization, showing a young sporophyte embryo. The embryo is covered by proliferated archegonial tissue. Following fertilization, cell division ceases in the notch meristem region of the gametophyte and cells enlarge. As the sporophyte continues to develop, the gametophyte eventually dies. Spore coat diameter, ca. 120 μm.

Close-up of hermaphroditic *C-Fern* gametophyte, ca. 3 days after fertilization, showing unfertilized (immature and mature) archegonia and young embryo developing within proliferated archegonial tissue. The remains of the old archegonial neck can be seen at the distal end of the embryo.

Figure 2.2 – Details of the gametophytic stage of the *Ceratopteris* life cycle. See text for details.

Q1   Explain why is it more difficult to select mutants from organisms that are diploid than from those that are haploid.

When a gametophyte with a non-wild-type phenotype is isolated, it is possible to generate a completely homozygous sporophyte from it in a single generation. This is something that is virtually impossible with higher plants and animals, but very desirable from the perspective of studies in genetics. Because the single-celled spore of *Ceratopteris* develops into a mature gametophyte via a process that involves only mitotic cell divisions, all of its cells (including eggs and sperm) will be genetically identical. Hermaphroditic gametophytes (including potential new mutants) can be easily isolated and self-fertilized, producing completely homozygous sporophytes. Such a sporopyte produces spores that are all genetically identical. If the plant is a true mutant, when it produces spores, all of them will have the mutant gene.

Cross-fertilizations can also be achieved by "watering" a mature hermaphrodite from one strain with a sperm solution from a second strain. A sperm solution with a high concentration of sperm can be produced from a culture with a high percentage of male gametophytes. Development of males can be induced by crowding gametophytes or by growing them on an antheridiogen-supplemented medium. Although self-fertilization is still possible under these conditions, crossing usually occurs because of the large number of sperm from the "male" strain. Also, genetic markers can be used to verify that cross-fertilization has occurred.

In this laboratory activity, you will investigate some of the characteristics that have been described here. You will observe some aspects of the life cycle of *Ceratopteris*, study a number of interesting mutant strains, and attempt to induce self- and cross-fertilization events using two strains of *Ceratopteris*.

PROCEDURE

General Considerations

You should work in pairs or small groups, as indicated by your instructor. Work together to accomplish the tasks described here, but make sure that each student sees for him- or herself the item or phenomenon that is described. You may also discuss the answers to the questions within your group; however, each student should formulate and write his or her own final response.

Observations

Dissecting and/or compound microscopes will be available for use in viewing various items/life stages of the *Ceratopteris* life cycle (dry spores, germinating spores, gametophyte cultures, sporophytes, etc.). Make general observations using available microscopes on each type of item. Also make the specific observations necessary to answer the following questions:

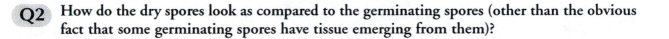

Q2 How do the dry spores look as compared to the germinating spores (other than the obvious fact that some germinating spores have tissue emerging from them)?

Q3 Observe a culture of "wild-type" gametophytes at a stage of growth from 10 to 14 days following soaking (DFS). What approximate percentage of the gametophytes appear to be males?

Q4 Sketch a region containing 10 to 20 cells from a gametophyte and a similar region from a sporophyte leaf. Describe the major differences between the surface cells of these two parts of the life cycle.

Observing Chemotaxis in *Ceratopteris* Sperm

Your instructor will provide you with a culture containing a high frequency of male gametophytes. You will also be given an aliquot of "sperm buffer," a pipette, a clean Petri dish, a concavity microscope slide, clean toothpicks (or dissecting probes), and several sharpened toothpicks previously soaked in a solution of L-malic acid. L-malic acid mimics the natural signal that comes from the archegonium and serves to attract the sperm. You should use the following procedure to generate a sperm broth from this culture and then observe how sperm respond to the L-malic acid added by dipping the toothpick into the sperm suspension.

Step 1: Add about 1 ml of sperm buffer to the Petri plate containing agar-solidified medium, but no spores or gametophytes.

Step 2: Using a clean toothpick or probe, scrape all the gametophytes except males from a region of the gametophyte culture and then scrape as many males as possible together into that area. Transfer as many males as possible into the sperm buffer in the other Petri plate. Do not transfer hermaphrodites, and avoid transferring injured males. Use your probe to mix/submerge the male gametophytes in the sperm buffer.

Q5 The chemotaxis response of sperm liberated from these males will be tested with L-malic acid. Why is it important not to transfer hermaphroditic gametophytes into this solution?

Step 3: Observe the transferred males for several minutes, using a dissecting microscope. Within about 3 to 4 minutes, there should be large numbers of rapidly moving sperm present in the sperm buffer. They should remain active for several more minutes. If you do not see any movement within the first several minutes, readjust the focus and lighting on the microscope.

Step 4: When active sperm are present, transfer one drop of sperm suspension to a concavity slide. Do *not* add a coverslip.

Step 5: Focus on the swimming sperm in the concavity slide with the low power objective (about 10X) on a compound microscope. Briefly touch the surface of the drop with the sharpened tip of a toothpick that was previously soaked with L-malic acid. Do not push the toothpick into the drop — just touch its surface. Observe what happens after the toothpick has been removed. To repeat your observations, mix the suspension with a clean probe or toothpick to disperse the L-malic acid, or obtain a fresh drop of suspension.

Q6 A. What do you think would happen if two toothpicks that contained the same amount and concentration of L-malic acid were touched to the opposite ends of the drop at exactly the same time? Explain your answer.

B. What do you think would happen if two toothpicks that contained L-malic acid at different concentrations (i.e., 15 mM and 30 mM) were touched to the opposite ends of the drop at exactly the same time? Explain your answer.

Working with Mutant Strains

During this portion of the laboratory, you will make observations on various "coded" cultures (a, b, c, etc.) of gametophytes and attempt to determine the type of mutation that exists in each. The following is a list of several mutant strains that are likely to be available:

dark germinator *(dkg1)* — The spores of this strain germinate in complete darkness. The gametophytes that grow from them will have an etiolated appearance, which means that they will be pale, long, and spindly (especially the cells nearest to the spore). This characteristic is not, strictly speaking, part of the mutant phenotype. Wild-type spores that are induced with light to begin germinating and then placed in darkness will also appear etiolated. The dark germinator mutants (and wild-type gametophytes grown in darkness) will also be deficient in chlorophyll and will appear very pale green. Unlike many higher plants, fern gametophytes produce some chlorophyll in darkness, but the amount produced is very small. You will be given a coded culture for this strain that will have just been removed from total darkness.

hermaphroditic *(her1)* — This strain has no male gametophytes.

highly male *(him1)* — This strain has a much higher-than-normal proportion of male gametophytes.

nonetiolated *(det30)* — To see this phenotype, spores must be moistened and *exposed to light to induce germination*, then grown in darkness. Under these conditions, gametophytes of this mutant strain do not grow like wild-type gametophytes, which become etiolated in darkness. Instead, nonetiolated gametophytes look somewhat more like normal, light-grown gametophytes, being somewhat broader at the base than etiolated gametophytes. These gametophytes will also be very pale green because they have been grown in darkness. You will be given a coded culture for this strain, which will have just been removed from total darkness.

pale *(pal)* — Gametophytes of this strain are light green compared to the wild type. For ease of identification, this mutant will be growing along with wild-type gametophytes as the F1 progeny of a cross between the wild and the pale strain.

Q7 What is the approximate ratio of wild to pale gametophytes in the culture with the F1 progeny of a cross between the wild and the pale strain? Can you propose a genetic model that fits the ratio?

polka dot (cp) — As the name implies, this mutant strain has the appearance of being polka dotted! Once you identify this strain, look at it more closely under the microscope and see whether you can determine *why* it looks polka dotted.

You should note that there are many other mutants of *Ceratopteris* that you did not encounter in today's lab. Among them are various physiological mutants, including one that is tolerant to high levels of salinity, several that are tolerant to various herbicides (e.g., paraquat and glyphosate), and also other developmental mutants, such as several that affect sperm motility like the mutants called "slow-mo" and "sleepy sperm" (your instructor may prepare a demonstration sperm suspension from one of these strains if it is available).

Q8 List the code designations of each culture that you observed, and indicate the name of the mutant phenotype exhibited in each case.

Self-Fertilization and Crossing Techniques for *Ceratopteris*

In this portion of the lab activity, you will attempt to initiate sporophyte formation on isolated hermaphrodites via self- and cross-fertilizations. The self-fertilizations should be successful as long as sufficient sperm is produced by watering an isolated hermaphrodite. Cross-fertilizations are likely to succeed, but you should verify this several days after the cross is made.

Many mutations can be used as markers to verify that cross-fertilization occurs. The following description will assume that you will be working with a phenotype that is particularly well suited for this purpose, the polka dot mutant.

The polka dot mutant produces an obvious, morphological phenotype that is caused by a recessive allele. When you use a polka dot gametophyte as the egg source (i.e., a polka dot hermaphrodite) and a wild-type strain as the sperm source, it is easy to determine whether the resulting sporophyte is produced by a successful cross-fertilization.

Q9 Why is it important that the strain used as egg source, and not the one used as the sperm source, has the polka dot mutation in the cross described above? What results should you expect if the cross described above is successful?

You should obtain an isolated hermaphrodite from your instructor (or prepare one yourself, if you are instructed to do so) and attempt to conduct a self- or cross-fertilization as assigned by your instructor. Label the plate containing your isolated gametophyte. After you have fertilized it, come back to determine whether your attempt was successful (1 to 2 weeks later should be sufficient in most cases).

Attempt to keep your materials as sterile as possible during the following manipulations. Otherwise, fungal contamination may occur and kill your isolated gametophyte and any sporophyte produced.

For self-fertilizations, water isolated hermaphrodites with sterile distilled water (about 0.1 to 0.5 ml). Use care to ensure that the gametophyte is completely submerged in the drop (move it with a sterile probe if necessary) and that no air bubbles block the sperms' access to the archegonia.

For cross-fertilizations, generate a sperm broth using sterile distilled water, and use this to water the isolated hermaphrodite. To generate the sperm broth, flood the plate containing a high density of wild-type males with sterile distilled water (a strain like the highly male strain, which is wild type for the polka dot phenotype, and can also be used as the sperm source). After a suspension of actively moving sperm has been produced, use drops of the suspension to water an isolated hermaphrodite, using the same techniques as those described earlier for self-fertilizations.

If you successfully produce sporophytes, and if your instructor is willing, they may be transferred to a greenhouse and maintained for long periods of time.

Ceratopteris is an excellent model organism for conducting research at an undergraduate level. You can learn more about this organism, including ideas for potential research projects, by visiting the C-Fern Web page (http://cfern.bio.utk.edu/index.html). This site is maintained by Les Hickok and Thomas Warne, the original promoters of *Ceratopteris* as a model system.

LAB ACTIVITY 3

Introduction to *Drosophila*
and Conducting Crosses

INTRODUCTION

The fruit fly, *Drosophila melanogaster*, is a common pest organism. However, the U.S. government and numerous research institutions expend millions of dollars each year to raise *Drosophila* . This is because *Drosophila* is one of the most important model organisms for studying the genetics of eukaryotic organisms. Since the early 1900s, when Thomas Hunt Morgan first began to study *Drosophila*, researchers have accumulated an amazing amount of information on the inheritance of various traits in this organism. Thousands of mutations have been characterized, and in 2000, *Drosophila* became the third eukaryotic organism for which the entire genome had been sequenced.

Although *Drosophila* is a relatively simple organism, it exhibits many physical and biochemical characteristics that exhibit variations (i.e., mutant phenotypes). This abundance of variety is one characteristic that has made this organism so fascinating to geneticists. Many of the other reasons it has become such a popular subject of study are related to the characteristics of its life cycle. First, it is relatively easy to distinguish males from females and to set up controlled crosses. Second, it is a very small organism and can be easily grown in large numbers in small spaces. Finally, it proceeds through its life cycle very rapidly, making it possible to conduct genetics experiments relatively quickly.

In this laboratory activity, you will become acquainted with the basic characteristics of *Drosophila* that make it useful for genetic studies. You will anesthetize flies and learn how to tell females from males, and you will learn to identify some of the mutant phenotypes. You will also learn the basic techniques needed to conduct crosses with this organism, and you will set up a simple cross. These procedures may be conducted during two or more scheduled lab periods, or your instructor may instruct you to come back to the lab on your own later to complete some of the steps (i.e., setting up a cross often requires several brief visits to the lab to obtain sufficient numbers of flies at appropriate stages).

PROCEDURES

Anesthetizing Flies

It is very important to learn to anesthetize flies properly. First, good anesthetization practices prevent the escape of flies. Second, good techniques are necessary to prevent flies from dying or becoming reproductively inviable, both of which are significant impediments to fly breeding.

The instructor will demonstrate this important technique during the lab. The following outline of the procedure assumes that you will be using ether for anesthetization. However, several other alternatives are available (CO_2, cold treatment, commercial anesthetics, etc.). If ether is used, ensure that proper ventilation is used, and avoid breathing the fumes. Excessive exposure to ether is hazardous to humans, as well as to flies! Also, do not use open flames when ether is in use, because ether is highly flammable.

To anesthetize flies with ether, you will need the following materials: cultures of flies, a funnel, ether, a transfer bottle (with a stopper), an etherizing bottle (with a stopper with cotton attached), and a reetherizer (a Petri plate lid with cotton attached to the inside).

To view the anesthetized flies, you will need a dissecting microscope, a white 3 x 5-inch card (to hold the flies), and a fine-tipped paintbrush, used for moving individual flies.

The following is an outline of the steps used to anesthetize flies with ether:

Step 1: Organize the necessary materials.

Obtain a culture with lots of adult flies and identify a working area with a soft surface (several paper towels, a soft-cover book, etc.) on which to "bang" the fly culture. It is necessary to "knock" the flies down inside the culture before transferring them (discussed later), and the soft surface deadens the banging sound and helps to prevent culture bottles from breaking.

Dampen the cotton in your etherizing bottle and reetherizer with ether; place the stopper for the etherizer into the bottle (this allows ether vapor to saturate the inside atmosphere of the etherizing bottle).

Place the funnel into the transfer bottle and ensure that the stopper for this bottle is within easy reach.

Step 2: Knock the adult flies down within the culture. By lightly banging the culture bottle on the soft surface, flying and crawling flies are knocked down from the top of the bottle. This is necessary so that they don't escape during transfer. If glass culture bottles are used, *be very careful not to break the bottles.* Check culture bottles for hairline fractures prior to this step; do not use bottles if such fractures are evident. If intact bottles are banged softly, they should not break.

Step 3: Transfer flies to the transfer bottle. The transfer bottle is used to hold adult flies just prior to anesthetization. Adult flies will be shaken into this bottle and then later transferred into the etherizing bottle. It is possible to shake adult flies directly into the etherizing bottle from the culture; however, the ether fumes are more damaging to eggs and larvae than to adults, so a transfer bottle is used if it is desirable to obtain additional healthy adults from the culture at a later time.

To move the flies from the culture bottle into the transfer bottle, continue lightly banging the culture bottle on the soft surface. While continuing to bang the culture bottle (don't forget to keep banging!), remove the stopper from the culture vessel. If you continue banging the culture, the flies will remain at the bottom — if you stop, they will escape! Once the stopper is removed from the culture vessel (keep banging), use your free hand to steady the transfer bottle containing the funnel. Next, use one quick, fluid motion to invert the culture vessel into the transfer bottle, and immediately begin banging the combined transfer bottle-funnel-culture bottle. This will knock flies from the culture bottle into the funnel and into the transfer bottle.

Once the flies are inside the transfer bottle, they will usually not be able to escape by flying up into the narrow end of the funnel. At this point, replace the stopper into the culture bottle and proceed with the next step.

Step 4: Transfer flies from the transfer bottle to the etherizer. Use essentially the same technique described in Step 3 to move flies between these two bottles. Begin by knocking the flies into the bottom of the transfer bottle. While the flies are being knocked into the bottom of the transfer bottle (keep knocking!), remove the stopper with the ether-soaked cotton from the anesthetizing bottle, and transfer the funnel to the etherizing bottle. Keep knocking. Use the technique described earlier to shake the flies into the etherizing bottle. As quickly as possible (while knocking the flies down inside the etherizing bottle), replace the funnel with the stopper. Do not allow the flies to remain in the etherizer for too long.

Step 5: Remove and observe flies. It is important to find the correct balance between too little and too much anesthetization. The flies should be sufficiently immobilized so that they will not stand up when the etherizing bottle is rotated and/or slightly knocked. However, too much anesthetization will reduce fertility or even kill the flies. It is better to underanesthetize the flies if you are uncertain, and then use the reetherizer to reanesthetize them later, if necessary. The reetherizer is used by ensuring that the cotton is dampened with ether and then using it to completely cover flies that are awakening.

Handle anesthetized flies on a white 3 x 5-inch card, and move them while viewing with a dissecting microscope by pushing them with a fine-tipped paintbrush.

Distinguishing Differences in the Sexes

To conduct specific crosses with *Drosophila*, it is very important to be able to distinguish the two sexes. Crosses are generally made by placing several males of one strain and several virgin females from a different strain together in a fresh culture vessel. If the gender of any of the flies used in such a cross is incorrectly assessed, the entire cross will be invalid. To ensure that you can correctly identify male and female flies, follow the instructions below carefully.

Obtain and anesthetize several wild-type flies, and find at least one male and one female. The following are the most obvious sexual characteristics that help to distinguish the two sexes of *Drosophila* (also see Figure 3.1):

1. The posterior of the male body is dark; the female abdomen appears striped.

2. The tip of the male abdomen is rounded, whereas the female abdomen is elongated.

3. The male abdomen has five segments, while the female abdomen has seven segments.

4. The male has sex combs on the distal surface of the basal tarsal joint of the first leg; the female does not.

5. The male sex organ is enclosed by a genital arch, while the female sex organ is covered by a vaginal plate.

6. On average, the male body is smaller than the female body (however, this is not an extremely reliable difference).

Verify for yourself that you can actually see the differences in the distinguishing characteristics. As you continue your work and study various non-wild-type phenotypes, check to be sure you can distinguish males from females when these phenotypic variations exist.

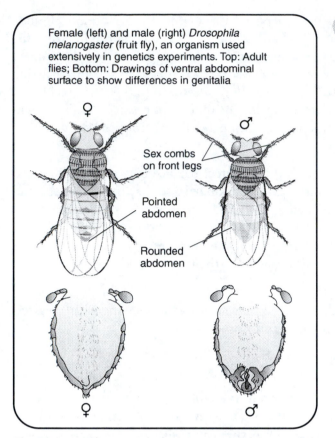

Female (left) and male (right) *Drosophila melanogaster* (fruit fly), an organism used extensively in genetics experiments. Top: Adult flies; Bottom: Drawings of ventral abdominal surface to show differences in genitalia

Sex combs on front legs

Pointed abdomen

Rounded abdomen

Figure 3.1 – The sexual characteristics of *Drosophila melanogaster*. See the text for descriptions of the distinguishing characteristics.

Q1 Which traits are likely to be the most reliable for distinguishing males from females and which are likely least reliable? Justify your choices by indicating how easy or how difficult it may be to observe the various traits in wild-type adult files. Also consider and describe how the traits might vary at different growth stages and for various phenotypes. (Note: You should probably wait to finalize your answer to this question until after you have viewed several flies with different mutant phenotypes.)

Distinguishing Different Mutant Phenotypes

In this part of the lab activity, you will become familiar with several mutant phenotypes of *Drosophila*, which your instructor has chosen for you to use. These flies are from the strains that you will be instructed to use for crosses at the end of this activity.

For this portion of the activity, you should work in small groups of several students. Obtain a collection of different coded vials from the instructor. Each vial contains flies that either are wild-type or exhibit one particular mutant phenotype. A list of the possible mutant phenotypes will be available in the lab. Your group should determine which phenotypes are present in each coded vial. Each member of the group should verify the conclusions made in this portion of the assignment. You can check unknown phenotypes against the flies available in stock cultures for verification. Be careful not to mix known flies with unknown flies! The instructor will have a list of codes and corresponding phenotypes; however, this list will not be accessed during the lab period.

Q2 List the codes for each of the vials that you worked with, and indicate the official designation for the phenotype of the strain present in each vial. Also provide a brief written description of the appearance of each phenotype you observed.

Conducting Crosses

In this part of the lab activity, you will be conducting a cross and analyzing the results. In each case, you will be working with strains that differ by only one gene (a monohybrid cross). Your goal will be to characterize the allele associated with the mutant phenotype (i.e., to determine whether it is dominant, recessive, etc.) and to determine whether it is associated with an autosomal gene or a sex-linked gene.

For this portion of the lab activity, you will work in small groups, and each group will be assigned to conduct a specific cross, using the wild-type and a particular mutant strain. Some steps of conducting the cross will have to be completed outside normal laboratory meeting times. Be sure that your group carefully coordinates the various aspects of conducting the cross.

Each group will be instructed to set up one or more replicates of the assigned cross in a reciprocal fashion. Doing the cross in reciprocal fashion means that each group will actually be conducting their cross in two ways. In one version of the cross, the females will be homozygous for the mutant allele and the males will be wild-type; in the other version, the phenotypes will be associated with flies of the opposite sex.

 Q3 Conducting crosses in reciprocal fashion is a typical aspect of genetic analysis. Why is it important to conduct reciprocal crosses — that is, under what circumstances might different results occur with reciprocal crosses? Explain your answer.

The *Drosophila* Life Cycle

Before you can conduct crosses, you must be familiar with the basic aspects of the fruit fly life cycle (see Figure 3.2). Several manipulations you will conduct must occur at specific stages of development.

The timing of the life cycle will vary somewhat at different temperatures. The life cycle speeds up slightly at higher temperatures; however, higher temperatures also promote fungal and/or bacterial growth (which can ruin cultures) and may cause variations in some phenotypes. If cultures are maintained at 21°C, mature flies can grow from eggs in about 14 days. All stages of development should be present in a healthy, mature culture. As you read the following description, you should try to identify the major, visible stages in such a culture.

The first several stages are not easily visible with the unaided eye or even a dissecting microscope. The tiny egg laid by the female usually hatches within 24 hours. The larva that emerges from the egg passes through three "instar" stages (which are separated by larval molts). The instar stages last from less than 24 hours to several days. At 21°C, the fly spends about 6 days in the egg/larval stages. The second and third instars are relatively large and can be seen with a dissecting microscope. The third instar stage is clearly visible with the unaided eye, especially when it begins the "wandering" stage, at which time it usually crawls up onto the culture vessel wall. The fly spends approximately the next 6 days in the pupal stage. It is while the fly exists as an immobile pupa that the incredible, internal changes of metamorphosis occur. At the end of the pupal stage, the adult fly (also called the *imago*) ecloses from the pupal case, and very quickly becomes capable of reproducing.

Drosophila Life Cycle

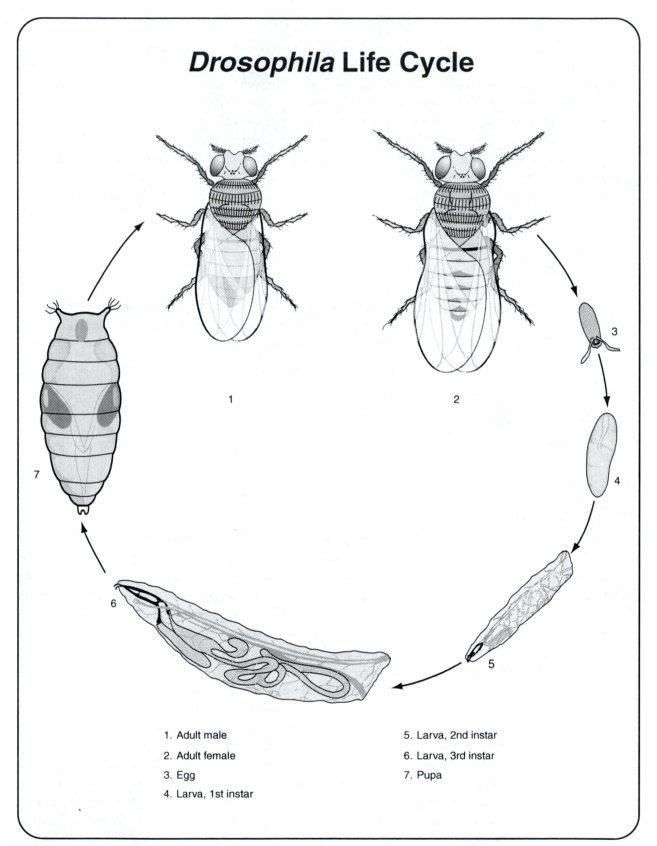

1. Adult male
2. Adult female
3. Egg
4. Larva, 1st instar

5. Larva, 2nd instar
6. Larva, 3rd instar
7. Pupa

Figure 3.2 – The life cycle of *Drosophila melanogaster*. See text for a detailed description.

Preparing Culture Vials

You will need fresh culture vials containing fly medium for crossing your P1 and F1 flies. Different kinds of culture media are available for this purpose. Your instructor may choose to provide premade culture media or have you prepare your own media (mixes with very simple preparation instructions are available). In either case, be sure to label your culture vessels with specific information about each cross you prepare (including your name, information about the cross, and date).

Preparing the P1 Crosses: Anesthetizing Flies and Selecting Virgin Females

The parental strains (P1 strains) are the homozygous flies that are initially crossed to obtain the "hybrid" offspring (called the first filial, or F1 generation). As indicated earlier, you will be conducting your cross in reciprocal fashion. Great care must be taken to correctly determine the sex of each fly used in the P1 generation. Furthermore, all of the females used must be virgins (discussed below).

The general technique for anesthetizing flies is the same as described earlier. However, you should be especially careful not to overanesthetize flies, since this can reduce fertility. When anesthetized flies are placed into culture vessels, care should be taken to ensure that they don't become stuck in wet culture medium. It is wise to place anesthetized flies onto the culture vessel wall in a stoppered vessel that is oriented on its side. This allows flies to revive on the dry wall surface, rather than the surface of the medium, which may be sticky.

Selecting flies to serve as the P1 generation must be done in several steps. Males can be isolated and added to a cross at any time (they can also be "saved" in a separate culture vessel for later use). However, when isolating females for a cross, steps must be taken to ensure that they are virgin females. Nonvirgin females that have been inseminated by males of the same strain will lay eggs for a long period of time, and these eggs will be homozygous for the genotype of the mother (i.e., not the result of the cross that you are trying to set up).

Since female flies cannot be inseminated during the first 8 to 12 hours of their adult life, virgins can be obtained by selecting females that have recently eclosed (i.e., "emerged") from their pupae (it is recommended that females be isolated as virgins no later than 8 hours after eclosure). To accomplish this, a stock culture with many late-stage pupae is used. All of the adult flies are first removed (the males can be used for the reciprocal cross), and newly eclosed flies are removed continuously (or at a period of time no later than 8 hours following clearing of adults). Newly eclosed females removed in this way can be used as virgins. Flies eclose most actively during the morning hours, so whenever possible, remove adults early in the day.

When you set up each cross (and each replicate), use about 5 to 10 males and about 5 to 10 virgin female flies for each culture. You can allow the P1 flies to remain in the culture and continue to mate for about 1 week to 10 days. However, you should watch your cultures closely, and *always remove P1 flies at or before the time when pupae first become visible* in the cultures. When the F1 flies eclose, remove a large number of these, and count and categorize them according to their phenotypes (including male vs. female as a phenotypic trait). Maintain a record of your results for future analysis.

Q4 Answer this question for your particular cross. If necessary, include separate responses for the two reciprocal crosses. Your instructor may ask you to hand in the answers to Parts A through E within a week or so (along with answers to the other questions in this lab) and then hand in an answer to Part F after you have collected the data from the cross.

A. What mutant phenotype are you using in this cross?

B. If the mutant allele in this cross is a dominant, autosomal allele, what phenotype(s) should be present in the F1 of your cross, and what fraction of the F1 should each phenotype represent? (Include "male" and "female" as part of your description for a given phenotype, only if it is necessary to do so.)

C. If the mutant allele in this cross is a recessive, autosomal allele, what phenotype(s) should be present in the F1 of your cross, and what fraction of the F1 should each phenotype represent? (Include "male" and "female" as part of your description for a given phenotype, only if it is necessary to do so.)

D. If the mutant allele in this cross is one that exhibits "lack of dominance" (i.e., codominance, or incomplete dominance) and is an autosomal allele, what phenotype(s) might be present in the F1 of your cross, and what fraction of the F1 should each phenotype represent? (Include "male" and "female" as part of your description for a given phenotype, only if it is necessary to do so.)

E. If the mutant allele in this cross is a sex-linked, recessive allele, what phenotype(s) should be present in the F1 of your cross, and what fraction of the F1 should each phenotype represent? (Include "male" and "female" as part of your description for a given phenotype, only if it is necessary to do so.)

F. What actual phenotype(s) were present in the F1, and what fraction of the F1 did each phenotype comprise? Based on these results, what type of allele (dominant, recessive, autosomal, etc.) does the mutant phenotype you studied seem to be associated with?

Preparing the F1 Crosses

Establishing crosses with F1s as the parents is simple compared to establishing P1 crosses. As indicated earlier, you must have removed the P1s at a time prior to when the F1 adults were first present (so that F1s were not able to mate with P1s). After you remove the P1s, you can collect and use F1s for the next cross without ensuring that the F1 females are virgins.

Q5 Why is it unnecessary that the F1 females be virgins prior to establishing the cross?

To set up the F1 crosses (remember to continue with reciprocal crosses), simply remove about 5 to 10 male F1s and 5 to 10 female F1s from the culture that previously contained the P1s, and transfer them to a fresh, labeled culture. You can make several replicates of the F1 crosses by conducting this manipulation for several days over the course of about 1 week after the F1s first eclose. Do not use flies from this culture that develop later than 1 week after the F1s eclose, because some of these flies could be the offspring of the F1s.

You can allow the F1 flies to remain in the culture and continue to mate for about 1 week to 10 days. However, you should watch your cultures closely and always remove F1 flies at or before the time when F2 pupae first become visible in the cultures. When the F2 flies eclose, remove a large number of these and count and categorize them according to their phenotypes (including male vs. female as a phenotypic trait). Maintain a record of your results for future analysis.

Q6 Answer this question for your particular cross. Include separate information for the two reciprocal crosses. Your instructor may ask you to hand in the answers to Parts A through E within a week or so and then hand in an answer to Part F after you have collected the data from the cross.

A. What phenotype(s) was present in the F1 flies in this cross?

B. If the mutant allele in this cross is a dominant, autosomal allele, what phenotype(s) should be present in the F2 of your cross, and what fraction of the F2 should each phenotype represent? (Include "male" and "female" as part of your description for a given phenotype, only if it is necessary to do so.)

C. If the mutant allele in this cross is a recessive, autosomal allele, what phenotype(s) should be present in the F2 of your cross, and what fraction of the F2 should each phenotype represent? (Include "male" and "female" as part of your description for a given phenotype, only if it is necessary to do so.)

D. If the mutant allele in this cross is one that exhibits "lack of dominance" (i.e., codominance, or incomplete dominance) and is an autosomal allele, what phenotype(s) might be present in the F2 of your cross, and what fraction of the F2 should each phenotype represent? (Include "male" and "female" as part of your description for a given phenotype only if it is necessary to do so.)

E. If the mutant allele in this cross is a sex-linked, recessive allele, what phenotype(s) should be present in the F2 of your cross, and what fraction of the F2 should each phenotype represent? (Include "male" and "female" as part of your description for a given phenotype, only if it is necessary to do so.)

F. What actual phenotype(s) were present in the F2, and what fraction of the F2 did each phenotype comprise? Based on these results, what type of allele (dominant, recessive, autosomal, etc.) does the mutant phenotype you studied seem to be associated with?

Q7 Compare the frequencies of phenotypes in your data (for each reciprocal cross) to the frequencies that would be predicted from a model based on your response to question 6F. Use chi-square analyses to compare the data to the ideal ratio(s). What are the chi-square values you obtained from these comparisons? What are the approximate p values associated with the chi-square values? Do these analyses support your assessment of the mode of inheritance you suggested in response to question 6F? (Show your calculations.)

The crosses you and your classmates conducted for this lab activity represent the "classical," Mendelian, approach to genetics analysis. The kind of knowledge gained using this approach is extremely important. It often forms a foundation for many types of genetic analyses that are more technically sophisticated. Another laboratory activity in this manual build on this foundational understanding of *Drosophila* genetics by introducing molecular techniques to study a genetic marker in this organism.

Mitosis and Meiosis:
Structure and Function of Eukaryotic Chromosomes

INTRODUCTION

In this lab activity, you will work with material from several organisms to investigate the processes of mitosis and meiosis, and the structure and function of eukaryotic chromosomes. The work you will be doing illustrates a few of the approaches used in a subdiscipline of genetics called *cytogenetics*. Cytogenetics is the study of the form and function of chromosomes. Geneticists who use cytogenetics techniques in their research do so to study a wide variety of phenomena. For example, medical researchers use a technique called *karyotype analysis* to study changes in chromosome number or structure that may cause specific diseases. Agricultural researchers use cytogenetics to study the chromosome makeup of crop and livestock strains. And understanding the basic chromosome makeup for a given organism has been the starting point for constructing genetic maps in all of the eukaryotic genome projects.

In this activity, you will study material from three different organisms because the material is readily available, easy to work with, and useful for illustrating specific features. You will be working with prepared slides of onion root tips to observe some features of mitosis and lily anthers to observe some aspects of meiosis. You will also be preparing chromosome "squashes" from the salivary glands of *Drosophila* larvae to observe giant polytene chromosomes. Polytene chromosomes are an unusual type of chromosome found in the salivary glands (and other organs) of the larva of many dipteran insects, including *Drosophila*. They are many times larger than the normal chromosomes of these species and are therefore convenient objects of microscopic investigation. Furthermore, stained polytene chromosomes exhibit characteristic banding patterns that can be correlated with genes on genetic maps in organisms such as *Drosophila*.

PROCEDURES

Studying Mitosis and Meiosis

A good understanding of the fundamental processes of mitosis and meiosis is essential for understanding how genes are passed from one generation to another. The following descriptions and instructions assume that students already have some understanding of these processes. Some questions will be asked that are not answered in the material presented here and for which answers are not obvious from observations of the slides. If you do not already know the answers to these questions, you should review a textbook that describes mitosis and meiosis in detail to answer them.

Also, the following material will not focus on many of the smaller details of mitosis and meiosis (although such details are important!). The main purpose of the following exercises is to help students review the major trends in mitosis and meiosis. Figure 4.1 is provided as a visual aide to help you understand these major trends. It is a highly formalized diagram, however, and shows only the major similarities and differences between mitosis and meiosis. Many of the stages described later are not shown in Figure 4.1. This has been done intentionally to encourage you to read the descriptions carefully and search for the stages on your prepared microscope slides.

In your study of mitosis and meiosis, you will look first at the major aspects of mitosis in the onion root tip. Then you will consider the major aspects of meiosis in material from the lily anther. Lastly, you will be asked to compare and contrast various aspects of the two processes.

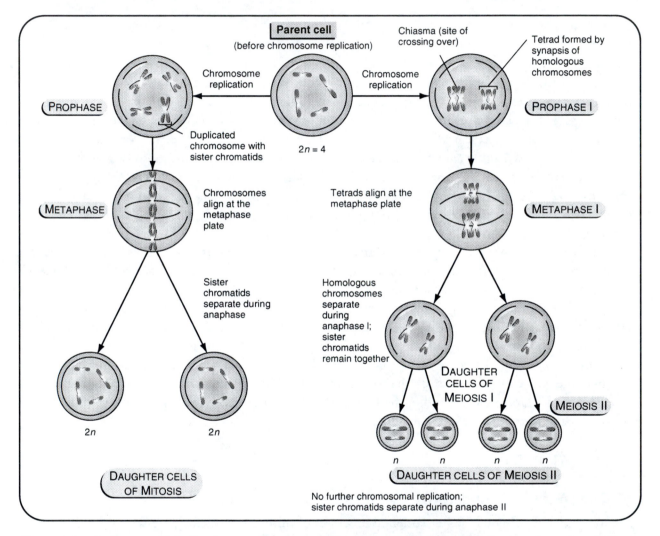

Figure 4.1 – A diagram of the fundamental processes of mitosis and meiosis. See the text for a description.

Mitosis in the Onion Root Tip

You will be looking at a long section of an onion root tip that contains many cells that were actively undergoing mitosis when the material was preserved. The best location in the root tip for finding cells that were actively dividing is just behind the root cap. The root cap is the most distal part of the root tip. The cells in the cap are much larger than the actively dividing cells just behind them (i.e., in what would be a more proximal location). As you search for actively dividing cells, you should not look too far away from the root cap, because cells just above the region of active cell division have stopped dividing and have begun to differentiate. The region of actively dividing cells is about equal in length to the length of the root cap. For each stage listed here, find a cell in the indicated stage, and answer the associated questions.

 A. Why would you not expect to see cells undergoing meiosis in the root tip? (The answer should seem obvious, so simply provide a short answer.)

B. Why is the root tip a better place to look for mitosis than another part of the plant — the leaf, for example? (Use the comparison between the root and the leaf in your answer to this question.)

Interphase

Technically, interphase is not an actual stage of mitosis or meiosis; however, it always happens immediately before either of these processes.

Q2 Why is interphase really not a stage of either type of cell division? Give an example of a situation in which this part of the cell's life is disconnected from both types of cell division.

During interphase, the DNA is not visible as discrete chromosomes. Also, dark structures called *nucleoli* (the sites where rRNA genes are transcribed) are present at this stage and may be visible if the material was sectioned in a manner that reveals them. Look for a cell in the interphase stage, and try to identify one with an evident nucleolus.

To answer the next question, you need to know what happens to the DNA during interphase before a cell goes through mitosis.

Q3 Assume that a gamete cell has 1X amount of DNA. You are probably used to thinking of gametes as having *n* amounts of DNA, but don't worry about the change in symbols used here — *n, 2n,* etc., are used to refer to ploidy levels (i.e., the number of sets of chromosomes); X is used to refer to amounts of DNA, irrespective of the ploidy level. If a gamete has 1X amount of DNA, how much DNA (1X, 2X, 3X, etc.) will be present in an onion root cell (a) at the beginning of interphase (just after telophase of a previous division) and (b) at the end of interphase (just before mitosis)?

Prophase

During prophase, the chromosomes condense and become visible under the microscope as discrete structures. They lie in scattered, unorganized fashion throughout the nucleus. Because the nuclei of the onion root cells are oriented randomly when thin sections are made from them, you can almost never find a "perfect" view of prophase stage chromosomes (remember, microscope sections are 2-D slices through 3-D structures). However, you should be able to see enough detail to recognize a prophase-stage nucleus. Identify one cell that appears to be in prophase.

Q4 How much DNA (1X, 2X, 3X, etc.) is present in an onion root cell during prophase?

Q5 A. If you could see a condensed, prophase-stage chromosome flattened out in two dimensions, it would have a very characteristic appearance. Draw a sketch of what such a chromosome would look like, and include it with your other answers.

B. Why does such a chromosome have the particular shape it does?

Metaphase

During metaphase, the chromosomes are lined up in the center of the cell. It is relatively easy to find nice-looking metaphase-stage cells in onion root cells because each cell has a "top" and a "bottom" (relative to the long axis of the root), and the metaphase plate will always form in the center part of the cell between the top and the bottom.

At this stage, the chromosomes will begin to attach to a structure called the *spindle apparatus* that is made of microtubules. In the next stage, the two sides of the spindle apparatus will contract pulling the attached chromosomes to opposite ends of the cell. Identify a cell that appears to be in metaphase. Try to find one in which the spindle apparatus is visible.

Anaphase

During anaphase, each side of the spindle apparatus contracts, pulling chromosomes to opposite ends of the cell.

 Q6 Find a cell that appears to be in anaphase and sketch it.

Q7 When the chromosomes separate during anaphase, what mechanism determines which chromosomes will go to one end of the cell, and which will go to the other end? Is this a random process or an organized one? To answer this question, you may wish to begin by describing the structure of the chromosomes at metaphase, explaining how the spindle fibers attach to them at this stage.

Telophase

During telophase, the chromosomes that were pulled to opposite ends of the cell during anaphase begin to form their own nucleus. Also, by the end of telophase, two new cells have formed. Identify a pair of cells that appear to be at some point in the telophase stage.

Q8 At the end of telophase, how much DNA (1X, 2X, 3X, etc.) does each new cell have?

Meiosis in the Lily Anther

The anther is the structure in the flower where pollen grains are produced. Young anthers from immature flower buds contain cells that have not yet initiated meiosis. Older anthers contain cells in different stages of meiosis, and the most mature anthers contain pollen grains. To see the various stages of meiosis leading up to the production of pollen grains (also called *microsporogenesis* because the cells produced at the end of meiosis are known as *microspores*), you will need to observe several slides. Your instructor will tell you which slides are available and what stages should be visible on the differently labeled slides.

The slides you will view are cross sections through anthers at various stages of maturity. Each anther has four chambers (each chamber is considered a microsporangium) that contain both nutritive cells (tapetum toward the outside of the microsporangium) and cells that are destined to become pollen grains (called *microspore mother cells*; surrounded by the tapetum). You will be instructed to observe material for some, but not all, stages in the following description.

Interphase

You will find interphase-stage cells on the slides containing material from the most immature anthers. Like the interphase cells of mitosis, which you viewed in the onion root tip, these will have diffuse DNA in their nucleus and nucleoli should also be present (though these will not be visible in each such cell). Identify a cell in the interphase stage.

Q9 How much DNA (1X, 2X, 3X, etc.) will be present in a microspore mother cell (a) at the beginning of interphase (just after telophase of the last mitotic division) and (b) at the end of interphase (just before meiosis)?

Prophase I

A difference between meiosis and mitosis is that meiosis has two sets of divisions. Therefore, each stage after interphase is designated as either a first division stage (I) or a second division stage (II). Prophase I is superficially similar to prophase of mitosis, but there is one major difference: during prophase I of meiosis, homologous chromosomes pair with each other. This pairing is vastly significant for the function of meiosis; before going any further, make sure you understand why the pairing of homologous chromosomes is so important in meiosis.

If time allows, view a slide that contains this stage of meiosis and try to identify a prophase I stage cell. You probably won't be able to tell that the chromosomes are paired.

Q10 What are homologous chromosomes?

Q11 If you could see a condensed pair of prophase I-stage chromosomes flattened out in two dimensions, you'd notice that they would have a very characteristic appearance. Draw a sketch of what such a pair of chromosomes would look like, and include it with your other answers.

Q12 Why is it so significant for the function of meiosis that the homologous chromosomes pair during prophase I of meiosis?

Metaphase I

Like metaphase of mitosis, chromosomes are lined up in the middle of the cell (however, there is one significant difference) attached to the spindle apparatus. It will be harder to find a distinct metaphase-stage cell in the anther cross section than it was in the onion root long section, because the cells are randomly oriented. If time allows, identify such a cell.

Q13 What is the difference between the way the chromosomes line up during metaphase I of meiosis as compared to metaphase of mitosis?

Anaphase I

Chromosomes are being pulled to opposite ends of the cell by the condensing spindle fibers.

Q14 How is anaphase I of meiosis different from anaphase of mitosis? Is the separation of chromosomes into the two ends of the cell random, or is it specific? If it is specific, describe what makes it specific.

Telophase I

The separation of chromosomes is completed and cytokinesis occurs to produce two new cells.

Q15 At the end of telophase I (after cytokinesis), how much DNA (lX, 2X, 3X, etc.) does each new cell have?

Prophase II

The chromosomes become evident again as thickened structures that are spread out from each other and located in a central nucleus. At this stage, the two cells resulting from the completion of cytokinesis remain joined together. Identify a pair of cells that each appear to be in prophase II.

Q16 Sketch a pair of cells in prophase II.

Metaphase II

In each of the attached cells, the chromosomes line up at the center of the cell. If time permits, identify such a pair of cells.

Q17 A. In what way would metaphase II in lily anther cells be different from metaphase of mitosis in lily root tips?

B. In what way would metaphase II in lily anther cells be similar to metaphase of mitosis in lily root tips?

Anaphase II

In both attached cells, chromosomes are pulled in opposite directions into the two ends of each cell. This is very beautiful to see if you can find a pair of cells in the correct orientation! Take some time to try to identify a pair of cells like this.

Telophase II

The separation of chromosomes is completed, and cytokinesis occurs to produce a total of four cells from the two attached cells.

Q18 At the end of telophase II, how much DNA (1X, 2X, 3X, etc.) does each of the four cells contain?

Comparing Mitosis and Meiosis

To summarize your review of the major trends in mitosis and meiosis, complete the following grid and turn a copy of it in with your assignment.

Process or Stage of Mitosis and Meiosis	How much DNA (1X, 2X, etc.) is present **per cell** at the beginning of each process?	How much DNA (1X, 2X, etc.) is present **per cell** at the end of each process?	Place an X in the box for each stage during which DNA is replicated.	Place an X in the box for each stage during which homologous chromosomes are paired.
Interphase before mitosis				
Mitosis— prophase				
Mitosis— metaphase				
Mitosis— anaphase				
Mitosis— telophase				
Interphase before meiosis				
Meiosis— prophase I				
Meiosis— metaphase I				
Meiosis— anaphase I				
Meiosis— telophase I				
Meiosis— prophase II				
Meiosis— metaphase II				
Meiosis— anaphase II				
Meiosis— telophase II				

Observing Polytene Chromosomes

As indicated earlier, polytene chromosomes are an unusual type of chromosome found in the salivary glands of fruit fly larvae. They are "giant" chromosomes that can be easily visualized using a standard laboratory microscope. Also, when stained, they exhibit characteristic banding patterns that can be correlated with genes on genetic maps in organisms such as *Drosophila*. These chromosomes exhibit the unusual phenomenon of somatic pairing. This means that each "individual" polytene chromosome actually represents a pair of homologous chromosomes. Somatic pairing increases the size of polytene chromosomes somewhat; however, the major increase in size comes when these undergo several rounds of DNA replication without the normal, accompanying mitoses. This results in structures composed of up about 1,000 strands of DNA lying side by side, perfectly aligned on a point-by-point basis.

The existence of polytene chromosomes represents one mechanism of gene amplification. These structures are found only in specialized organs such as the salivary glands. Such organs are often the site of the production of very large amounts of a small number of substances. Polytene chromosomes represent a mechanism for increasing the numbers of genes by about 1,000 times in individual cells. Although all of the genes on all of the chromosomes in such cells are thus amplified, only certain genes in each cell type (i.e., salivary cells, etc.) are active. The evidence of this feature can be seen in the phenomenon of "chromosome puffs." Puffs on polytene chromosomes are areas of gene activity where the chromosome structure becomes diffuse and spreads out around the arm of the chromosome during transcription. Puffs occur in known patterns on the different chromosomes, and these patterns vary among different cell types. They also appear in specific fashion after larvae have been exposed to various environmental and/or chemical treatments. These variations in patterns of chromosome puffing suggest that certain chromosome bands represent specific genes that are being preferentially expressed.

PROCEDURE FOR PREPARING

Drosophila Polytene Chromosomes

Step 1: Obtain a third-instar *Drosophila* larva that has moved onto the wall of a culture vial and begun to slow down its movements. Such a larva is preparing for pupation and is at the ideal stage for this investigation.

Step 2: Place the larva onto a small drop of physiological saline on a glass slide. Using a dissecting microscope, remove the salivary glands using the following technique: Locate the head of the larva (smaller end), and pin it down to the glass with the point of a dissecting needle. Place another dissecting needle onto the body of the larva, about one-third of the body length away from the head. Pressing both needles firmly, pull the body of the larva away from the head. If this is done carefully, the internal organs, including the salivary glands, digestive tract, and fat bodies, should remain attached to the head (see Figure 4.2).

Step 3: Separate out the two transparent salivary glands. They will be closely associated with the two larger fat bodies that are the most opaque structures attached to the head. Be careful always to keep the salivary glands moist. Remove the other organs and body parts to the extent that it is possible to do so.

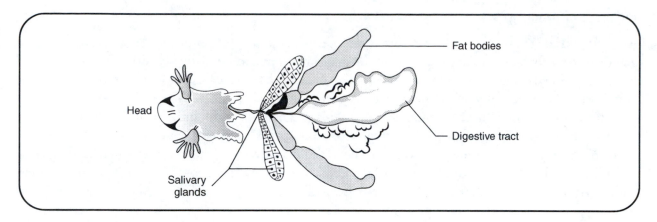

Figure 4.2 – A diagram showing the internal organs, including the salivary glands, digestive tract, and fat bodies, of a third-instar *Drosophila* larva. See the text for description.

Step 4: Place a small drop of aceto-orcein stain onto the glass slide nearby (but not touching) the drop of saline. Carefully lift the salivary glands out of the saline, and place them into the drop of aceto-orcein, then wipe up the saline solution on the slide with a piece of absorbent paper. Allow the glands to remain in the stain for 6 to 7 minutes before proceeding to the next step (experience may show that longer or shorter staining times produce better results). Add stain during the staining period if the drop begins to evaporate.

Step 5: After the staining period, place a coverslip over the stained tissue, as if you were making a standard wet mount. Then place a piece of paper towel over the slide and coverslip and carefully press directly down onto the coverslip with your thumb. This should be done with substantial pressure, but use care not to break the slide and to ensure that the coverslip does not move sideways during this procedure (this results in "rolled-up" nuclei and chromosomes).

Step 6: Remove the paper towel and add more stain if necessary. Viewing the squashed preparation, see whether you can find the extended polytene chromosomes. If the cells were not effectively squashed by the previous procedure, try applying pressure with your thumb again. You may need to make several preparations before you are completely successful, so don't give up easily.

Q19 Make a drawing of the terminal portion of one arm of a polytene chromosome. Include at least 8 to 10 bands on your drawing, and be careful to sketch these bands as accurately as possible (indicating the relative width, darkness, etc., for each band). Also, compare your drawing to published observations of *Drosophila* polytene chromosomes. Attempt to identify which chromosome (out of the four in the *Drosophila* karyotype) it was that you observed and illustrated. If your instructor chose to have you work with *D. virilis* (which is somewhat easier to use because of its larger size), the karyotype will be different from that of *D. melanogaster*. If you worked with *D. virilis*, indicate how many chromosomes you think are present.

L A B   A C T I V I T Y   5

Arabidopsis thaliana:
Rise to Fame in Plant Genetics

INTRODUCTION

At the beginning of the twenth century, *Arabidopsis thaliana* (Figure 5.1) was known only as an inconspicuous weed. By the end of the 20th century, it became the major model system in plant genetics. This rise to scientific world fame suggests something about how geneticists choose a model organism. Mendel's peas were the first model system in genetics, and after the rediscovery of his work in 1900, other agricultural plants, such as maize (corn) and tobacco, became major model systems. These organisms have some qualities that make them well suited as genetic models. For example, they are relatively easy to grow and maintain, and in each case, a single plant can produce hundreds or even thousands of seeds in one growing season. However, one of the main reasons that plants such as maize and tobacco became major model systems was strictly economic — they are important crop species, and funding agencies could easily see the advantage of experiments using them. This distinction of economic importance has become less significant in recent decades.

Arabidopsis is a tiny plant in the Brassica family, which also contains plants such as mustard, cabbage, and broccoli. However, unlike some of its close relatives and unlike the model systems of maize and tobacco, *Arabidopsis* has virtually no economic value whatsoever. But it does have several features that make it ideal as a model organism. *Arabidopsis* has been called the *Drosophila* of the plant world, and like *Drosophila,* it would just be considered another pest species if it weren't for some of its unique features. *Arabidopsis* is very small (compared to most plants, just a few centimeters high at maturity), it is easy to grow and breed, it matures quickly (in about 6 to 8 weeks), it can produce vast numbers of seeds (as many as 10,000 per plant), and it has a relatively small genome (with only five pairs of chromosomes).

Arabidopsis was first recommended to the scientific community as a model organism by Friedrich Laibach in 1943. However, for many years, only a small group, including Laibach and his students, utilized this organism. However, during the last several decades, *Arabidopsis* has become the major model system among plants. It is currently studied by hundreds of major laboratories, and in 1990, it became the focus of the first plant genome project (which was completed at the end of 2000).

As with other model organisms in genetics, there are hundreds of known mutants of *Arabidopsis.* In this lab activity, you will observe several mutant strains of this organism and compare them with the wild type. You will then have the opportunity to identify and isolate plants that may contain new mutations. To find these potential new mutant plants, you will be visually screening large numbers of seedlings derived from a mutagenesis experiment (which used ethyl methane sulfonate as a mutagen). The seeds that produced these plants are called M2 *seeds* (mutagenized generation 2) because

This lab activity was inspired in part by a poster presented by Jonathan Monroe to the Council on Undergraduate Research in 1994. The poster, which describes his use of *Arabidopsis* for a plant physiology course, is located at http://csm.jmu.edu/biology/courses/bio455_555/atlab/poster1.html

they represent a large batch of seeds derived from plants grown from mutagenized seeds (M1 seeds). Because most mutations are recessive and because *Arabidopsis* is a diploid organism, most mutations are generated by mutagenesis would never be observed in the M1 generation. Therefore, it is a common practice to allow M1 plants to self-fertilize and screen M2 seeds for possible new mutations. If your class is able to isolate several putative mutants, your instructor may have you grow these plants and self-fertilize them to produce seeds so that the mutation can be confirmed.

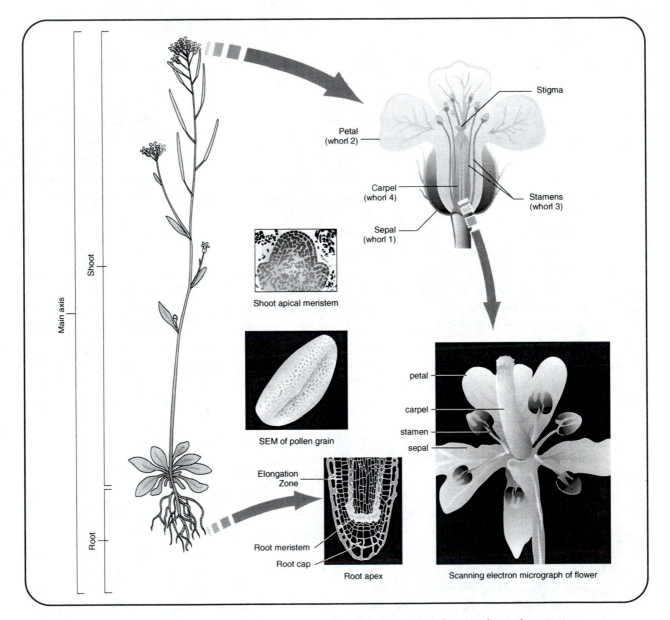

Figure 5.1 – Diagram of *Arabidopsis thaliana* root tips and flowers. See the text for a description.

PROCEDURE

Observing Wild-Type *Arabidopsis* Plants

Obtain several wild-type plants at different stages of development, observe the features described here, and answer the questions related to these features. After you become familiar with some traits of the wild-type strain, you will be given plants from several mutant strains and asked to identify the mutant traits. In most cases, these mutant traits will be associated with one of the features that you will be directed to observe in the wild-type strain during this part of the activity.

General Shape, Color, and Appearance

Begin your observations of the wild-type strain by noticing the general features of its appearance. First, notice the basic arrangement of the parts of the plant itself. Observe the general shape and arrangement of parts in both young seedlings and in more mature plants. Notice that the youngest seedlings are very simple in appearance, having little more than a primary root (with root hairs), a hypocotyl (the portion of the stem below the cotyledons), and two obvious cotyledons (seed leaves). There are also less-obvious leaf primordia (precursors to future leaves) nestled between the two cotyledons. In slightly older seedlings, both cotyledons and young vegetative leaves will be obvious. The mature plant has a much more complex organization. The oldest leaves on mature plants form a "basal rosette" just above the soil from which a comparatively long, branched stalk emerges. Notice the placement of branches, leaves, and flowers on the mature plant.

Q1 Sketch the appearance of a young wild-type seedling, labeling the structures noted in the prior section.

Q2 Describe the general morphology of the more-mature plant you observed. In your answer, compare the appearance and arrangement of the basal leaves with those features of the upper leaves. Also describe the branching pattern and the flowers' position and arrangement.

Q3 Look closely at the surface of the leaves and stems of the seedlings and more-mature plants. How would you describe the color and the sheen (i.e., the brightness or shininess) of these tissues? Is the coloration and sheen of the plant uniform over the entire surface of a given plant, or does it vary?

Trichome Morphology

Now, using a dissecting or compound microscope, look carefully at the surfaces of the leaves and stems of the wild-type plants. Notice that tiny hairlike structures (called *trichomes*) are present on some of these surfaces. Many of the mutant strains of *Arabidopsis* have unusual, or even absent, trichomes. After looking carefully for trichomes on the various surfaces of seedlings and more-mature plants, answer the following questions:

Q4 Were trichomes present on both the seedlings and the more-mature plants that you observed? Which parts of each of these plants had trichomes, and which did not?

Q5 There are several different types of trichomes on wild-type *Arabidopsis* plants. What kinds of variations did you observe in trichome shape, and where were the different kinds of trichomes located?

Flower Organization

Flowers are complex reproductive structures that are usually very characteristic in their organization from species to species. In general, flowers are organized with concentric whorls of flower parts. From the outside in, the whorls of parts include the leaflike sepals; the more showy petals the male reproductive structures, called *stamens* (each with a filament and an anther); and the female structures, called *pistils* (each composed of a basal ovary, a long style, and a stigma where pollen is received). Using a dissecting microscope, carefully examine a typical flower from a mature *Arabidopsis* plant, and answer the following question:

Q6　How many sepals, petals, stamens, and pistils does a wild-type *Arabidopsis* flower contain?

Observing Various Mutant Strains of *Arabidopsis*

Your instructor will provide plants representing several different types of mutations. Some will be very young seedlings, and others will be more mature plants (since some mutations will be more obvious at certain stages). These plants will be in containers labeled with "code" designations, but not the names of the mutant phenotypes. You should carefully observe these plants, using dissecting and/or compound microscopes as needed, and determine which phenotypes are represented by each labeled culture.

The following are descriptions of some of the possible phenotypes that may be present in the strains available in your lab. Your instructor may provide more information regarding which of the following mutants may or may not be present among the plants provided in your lab. You may also be provided with names and descriptions of other mutant phenotypes besides those listed here.

albina — These plants lack normal chlorophyll and are therefore cream colored. They only live a short while (as seedlings) and must be maintained in heterozygous stocks. If your instructor has provided some of these plants for you to observe, they may be present as a few seedlings in a mixture of F2 seedlings (including lots of green ones).

apetala — Several types of apetala mutants exist. They are homeotic mutations in which some part of the flower (often the sepal or petal) is absent or is transformed into another part of the flower, or into a leaflike structure.

chlorina — Due to a chlorophyll deficiency, these plants have yellowish-green shoots, instead of the normal darker green.

chloroplast mutator — These plants have sectors of green, yellow, and white variegation.

constitutively photomorphogenic — Plants with this category of mutation respond abnormally to light signals that control growth and development. When grown in the light, they accumulate purplish anthocyanin pigments in the cotyledons. When grown in darkness, they do not exhibit the normal, dark-grown, etiolated (i.e., "spindly") appearance. These plants are maintained in heterozygous stocks and may be mixed with normal plants as F2 progeny.

distorted trichomes — The trichomes on these plants are short, bent, and clublike.

dwarf — As the name implies, these plants are short (even for *Arabidopsis!*).

eceriferum — The stems of these plants are bright green and glossy, due to defects in the epicuticular wax layer.

erecta — These plants have short petioles and siliques (seed pods) and compact inflorescences.

floral mutant — The flowers of these plants look normal on the outside (sepals, petals, and stamens) but may have additional numbers of stamens and/or organs that appear to be fusions of stamens and carpels, and/or they may be completely or partially lacking in female reproductive structures.

glabra — There are several glabra mutations. All either have defective trichomes (i.e., unbranched) or are missing the trichomes, or the trichomes have been reduced in number on certain parts of the plant.

gnom — Seedlings with this mutation may lack a root and cotyledons, and they are cone or ball shaped. Obviously, this is a lethal mutation, and this gene must be maintained in a heterozygous stock.

lepida — This is another dwarf mutant. This strain is distinguished by having round, dark leaves.

root hair defective — The root hairs on these mutants may be short and wavy, branched, and/or short and bulging.

unusual floral organs — This floral mutation affects the second and/or third whorl of floral organs (petals and stamens), causing them to be absent in some cases or transformed into sepal or carpel-like organs.

yellow inflorescence — Unlike wild-type strains that have white flowers, this strain has yellow flowers.

The preceding list is just a small sampling of the hundreds of *Arabidopsis* mutants that have been discovered. By characterizing many of these odd phenotypes, investigators have learned some of the secrets of what controls normal growth and development in this tiny plant and in other plant systems as well.

Q7 List the code designations of each type of plant you observed and indicate the name of the mutant phenotype exhibited in each case.

Searching for New Mutations in *Arabidopsis*

Now that you have become familiar with the appearance of normal, wild-type *Arabidopsis* plants and have seen several mutant strains, you will have the chance to search for new mutations. Your instructor will provide you with one to several Petri plates with agar-solidified medium containing hundreds of small *Arabidopsis* seedlings. As was noted earlier, these are M2 seedlings that were derived from plants grown from mutagenized seeds (M1 plants). You should observe these with a dissecting microscope and look for mutant phenotypes like those described earlier (or anything else that looks unusual).

Although mutations occur very infrequently (even when a mutagen like EMS has been used), if you search carefully, there is still a good chance that you (or at least several people in your lab) will find a plant with a new mutation. Mutation rates vary with the amount and type of mutagen used, and they also vary for different genes. However, based on typical spontaneous mutation rates for eukaryotes (ranging from ca. 10^{-7} to ca. 10^{-4}), one might expect an "average" gene under optimally effective concentrations of EMS to become mutated in something like 1 out of 100,000 seeds. Finding one mutant plant out of 100,000 seedlings would be like searching for a needle in a haystack. However, when you visually screen the M2 *Arabidopsis* seedlings in this experiment, you will be looking for any visible mutant phenotype (not just one phenotype associated with one gene).

Presumably, there are thousands of genes (maybe even tens of thousands) that could produce a visible mutant phenotype that you could detect. This means that if you make careful observations and look for any apparently non-wild-type phenotype, you (and your classmates) should be able to find several possible mutants for every few thousand seedlings you screen. That's still a lot of seedlings to look at, but your chances of success are pretty high!

To ensure that you carefully view all the M2 seedlings on the Petri plate(s) you are given, you should draw a grid on the bottom of the plate(s) and view each square of the grid separately. Use a ruler and a permanent marker or a wax pencil to draw a grid with boxes that are about 1 cm square. As you carefully scan the seedlings in your plate(s), remove any with unusual phenotypes as possible mutants. Remove these by gently pulling them out of the agar-solidified medium with a pair of sterilized forceps. Using sterile technique, transfer possible mutants to small Petri plates containing fresh, agar-solidified growth medium, and label the Petri plates accordingly.

During the weeks to come, you should continue to observe any possible mutants that you isolated to determine whether the phenotype is stable during later development. If possible, you should transfer some or all of the putative mutants isolated by your class into soil, and keep them growing until they produce seeds (within about 6 to 8 weeks). At this point, the seeds can be harvested and grown to verify whether a heritable mutation is actually present. If one or more true mutations have been isolated, these could be further characterized by student researchers outside this class.

LAB ACTIVITY

Designing
PCR Primers

6

INTRODUCTION

The polymerase chain reaction (PCR) is a technique that has revolutionized many aspects of modern genetics. The technique was developed in 1986, earning its developers the Nobel Prize in 1993. It is a method of amplifying segments of DNA that generates millions of copies of a specified sequence in a period of hours. Since its development, PCR has replaced many of the traditional methods of molecular biology as a method of choice for various applications. These include aspects of DNA cloning (i.e., sequence amplification), DNA sequencing, and even the generation of specified mutations.

In this lab activity, you will use the World Wide Web to search a database for a gene sequence, and you will then use a Web-based program to design PCR primers for that sequence (an overview of the PCR process is presented shortly). As a follow-up exercise, your instructor may have you design primers for a gene of your choice. Such primers could actually be synthesized and tested if time and other resources permit (see Lab Activity 7, entitled "Using PCR to Find New Gene Sequences").

POLYMERASE CHAIN REACTION: BACKGROUND INFORMATION

In PCR, regions of DNA that are flanked by short, known sequences can be easily amplified. This provides enough copies of the amplified DNA so that it can be visualized on an electrophoresis gel and manipulated in various ways. Sometimes, the flanking sequences are known, but what lies between them is unknown. In these cases, PCR can be used to amplify a sequence that can be further studied to characterize it completely (this approach is illustrated in the last section of this lab activity entitled "Discovering New Gene Sequences Using PCR"). In other instances, PCR is used to manipulate DNA sequences that are already characterized. For example, PCR can be used to incorporate specific new mutations into genes carried on plasmids (PCR-based, site-directed mutagenesis).

In all cases, PCR amplifies DNA using a DNA polymerase that requires a primer. Primers are short, synthetic, oligonucleotides that are complimentary to DNA sequences that flank the region to be amplified. In most cases, the sequences of these primers are based on known sequences; however, one technique, called randomly amplified polymorphic DNA PCR, uses random primers. In addition to primers and DNA polymerase, PCR reactions must contain template DNA (the DNA to be amplified) and the DNA "building blocks" deoxynucleotide triphosphates (dNTPs, which include dATP, dTTP, dGTP, and dCTP). Furthermore, since the PCR process includes a high-temperature step, the DNA polymerase must be a thermostable polymerase.

Amplification of DNA by PCR is accomplished by a process known as *thermal cycling*. The PCR reaction (template DNA, primers, dNTPs, thermostable DNA polymerase, etc.) is combined in a tube

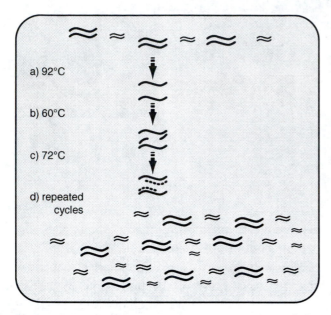

Figure 6.1 – A diagrammatic representation of the PCR process: (a) template DNA is denatured by high temperatures; (b) primers anneal to template DNA; (c) thermostable DNA polymerase synthesizes new DNA; (d) multiple cycles of the three temperatures result in an exponential increase in the sequence associated with the PCR process.

and placed into a DNA thermal cycler. The thermal cycler rapidly heats up and cools down, taking the reaction mixture through a series of temperature changes that occur as cycles. These cycles consist of three temperature stages that each accomplish different aspects of DNA amplification (see Figure 6.1). The first stage uses a high temperature (usually 92° or 94°C) that denatures the double-stranded DNA into single strands. This step is necessary before new DNA can be synthesized on the single strands. The second stage uses a lower temperature at which the synthetic primers recognize and form hydrogen bonds with the single-stranded DNA (this process is called *annealing*). The specific temperature of this stage depends on the base composition of the two primers (Figure 6.1 shows an annealing temperature of 60°C). Specific annealing temperatures can be calculated for any given PCR primer. The third stage uses a temperature that is optimum for the function of the DNA polymerase that synthesizes the new DNA (usually 72°C). The thermal cycler takes the reaction mixture through many cycles (often as many as 35) of these three temperature stages, essentially doubling the sequence flanked by the primers with each cycle. This process results in an exponential increase in the number of copies of the sequence, so that millions of copies of the PCR product are present at the end of the process. This amplified DNA is so concentrated that it can be seen as a discrete band when the DNA is visualized on an electrophoresis gel.

It is interesting to note that practical PCR was not possible until a thermostable DNA polymerase was isolated (one that would not be denatured by the 92° to 94°C DNA denaturation step). Such an enzyme (now called *Taq* polymerase) was first isolated from the bacterium *Thermus aquaticus* (= *Taq*) that thrives in hot springs, where most living things cannot survive.

PROCEDURE

The following steps will direct you through the process of generating two pairs of primers for the CRHB2 gene from the model plant system *Ceratopteris richardii*. *Ceratopteris* is a fern that is a relatively "new" model organism with great promise for studies in genetics (see the C-Fern page at http://www.c-fern.org for more information). In this exercise, each pair of primers you design will be specific for a particular region of the gene (designated as either region 1 or 2). Regions 1 and 2 were arbitrarily chosen to include the sequences from base number 50 to 150 (region 1) and from 500 and 600 (region 2). You will be given step-by-step instructions (to ensure success) for designing a primer pair for region 1 and less-specific instructions for designing primers for region 2.

Step 1: Go to the "Entrez Nucleotide" search page of the National Center for Biotechnology Information (NCBI) website as described below. The NCBI homepage provides links to many useful tools and databases. You will only be using a small portion of the resources available there for this activity. However, you may wish to surf the NCBI homepage a bit when you visit it to see what's available there. To obtain the sequence for the *Ceratopteris* CRHB2 gene, you will be using the "Entrez Nucleotide" search page. To get to this page, go to the NCBI homepage (www.ncbi.nlm.nih.gov/).

After arriving at the NCBI homepage, click the words "Entrez Home" in the column on the right. This will take you to a page where you can choose from several links, including one labeled "Core Nucleotides"; select this link. This will take you to a page entitled "Entrez Nucleotide."

Step 2: Searching for *Ceratopteris* gene sequences — In the search box, type in the name "Ceratopteris" (without the quotes) and click the Go button. When the search engine has completed the database search, you will see the results of the search, including an indication of the number of documents that were found that are related to *Ceratopteris*. Since *Ceratopteris* is a relatively "new" model organism, the number of sequenced genes is somewhat lower than would be the case for an organism like *Arabidopsis* (try the same search using "Arabidopsis" as the search term, and see how many records are listed).

Q1 How many sequences did you find when you searched using Ceratopteris as a keyword?

Step 3: Viewing information about the CRHB2 sequence — Now go back to the "Entrez Nucleotide" search box, type in the term CRHB2, and click the Go button. When the search engine has completed its search, click the link for the citation it finds, and you will see a page with information about this gene and the individuals who have worked on its sequence.

Q2 What is (are) the name(s) of the author(s) who published an article describing this sequence?

Q3 According to the article title in the GenBank report, what kind of gene is CRHB2?

Step 4: Copying the CRHB2 gene sequence — At the top of the page, you will see a pull-down menu labeled "display." Use the pull-down menu on that box to select "FASTA." When "FASTA" is selected this will take you to a page containing the nucleotide sequence for the CRHB2 gene in a form that can be copied and manipulated. Using the mouse or the keyboard, highlight the complete gene sequence (only the nucleotides), and hit the "copy" command in the browser menu. Once the sequence is copied to the computer's clipboard, you will access another Web page to work with the sequence to generate PCR primers.

Stop 5: Generating PCR primers for the CRHB2 gene — In this step, you will be using a Web-based program that designs primers based on the parameters you designate and other factors.

After you have supplied some specifications, the program uses various algorithms to generate the best possible primers that fit your specification. Various factors are involved in generating good primers. Ideally, they should only bind to the template DNA in the regions for which they were designed. They should also be composed of sequences that will not form internal hydrogen bonds or bonds between the two primers.

To access a Web page with a primer design program, go to the following Web address:

> Primer3 (a Primer Design page):
> http://frodo.wi.mit.edu/

When you arrive at this page, paste the gene sequence into the box at the top of the page. Next you will set some parameters that allow you to modify the primer selection process. As you set these parameters, click the associated links so that you can learn what types of specifications you're setting. If no setting is given in the following list for a particular parameter, leave the default setting as it is.

Sequence Id: Supply a name of your choice.

Target: 100,1

Product size range: delete the numbers given as possible default settings and enter 60-90

Primer size: minimum, 18; optimum, 18; maximum, 18

After entering the information, scroll down the page and click the "pick primers" icon, and the sequences for an optimum primer pair will be generated (a number of alternative primer pairs will also be listed following these). You will also see a representation of the gene with arrows indicating where the primers will associate with the gene sequence. Notice from this image and also from the information regarding the starting point for each primer that the PCR product should represent a portion of the gene located in region 1 (as described earlier). This result occurred because you entered appropriate parameters (e.g., target and product size) to obtain primers that should produce a product in region 1.

Q4 How big will the PCR product be if the optimum primers are used?

Q5 What are the sequences of the two optimum primers that were generated?

Now go back to the Primer3 input page and try modifying some of the specified parameters to see how it affects the size and placement of the PCR product that would be generated. After you are familiar with how the program works, generate a primer pair according to the following specifications. The primer pair you generate next should produce a PCR product that will be less than 80 base pairs long and located within region 2, as described earlier. The PCR primers themselves may be any size from 18 to 27 bases long.

Print the portion of the Primer3 output page that shows the optimum primer pair that you generate according the specifications given earlier, and attach the printout to your lab assignment when you hand it in.

Q6 The sequence for CRHB2 in the GenBank database was generated from mRNA (i.e., the sequence is equivalent to a cDNA sequence). If you wanted to use the primers you generated in this exercise to amplify sequences from genomic DNA (instead of cDNA), what other information would be beneficial to know about the gene sequence for CRHB2 before using these primers in an experiment? *Hint*: Sequences in eukaryotic mRNAs are generally quite different from genomic sequences because of a phenomenon known as posttranscriptional modification. In answering question 6, think about what these differences are and which of them might cause unexpected changes in the predicted size of your PCR product.

DISCOVERING NEW GENE SEQUENCES USING PCR

Your instructor may ask you to repeat the preceding process using a gene of your choice (or one assigned to you) as a follow-up activity. If time and resources permit, your primers (or those designed by another classmate) may actually be synthesized and tested, as described in a separate lab activity. If the gene and the primers are chosen carefully, you may be able to discover a new gene sequence (in a different organism) using these primers. The following are some tips that will help you design your primers for this purpose.

Choosing an Organism to Study

The organism for which you design new primers should be one for which relatively few genes have been sequenced. For example, as you saw earlier, far fewer genes from *Ceratopteris* have been sequenced than from *Arabidopsis*. Similarly, while the entire genome of *Drosophila melanogaster* has now been sequenced, very few genes from ant species have been described. To see what kinds of genes have been sequenced for a given organism, just run an "Entrez Nucleotides" search for that organism in the NCBI database. In a separate lab exercise, instructions are given for extracting DNA from plants and several small insect species, so that PCR primers can be tested using the DNA of these organisms.

Choosing a Gene to Study

If you hope to discover a new sequence, obviously you must focus on a gene that hasn't been sequenced in your organism (check GenBank). To design primers, you'll need to work from a known gene that has been sequenced in a related organism. For example, if you want to find a new sequence in *Ceratopteris*, start with a sequence from another plant and maybe even another fern. You'll have a better chance of picking primers that will work if you select a gene that is known to be highly conserved. Many genes that serve essential roles are highly conserved, meaning that they are structurally more similar between species than are other genes. For example, ribosomal genes, calmodulins, tubulins, and parts of most homeotic genes are known to be highly conserved. Finally, consider the source of the sequence information that you use to design your primers — is it from genomic DNA or cDNA, and what are the implications (i.e., see question 6)?

Testing Your Primers

If you are able to have your primers synthesized, you can isolate the DNA from your organism of choice and conduct PCR using these primers (see Lab Activity 7, entitled "Using PCR to Find New Gene Sequences"). If your primers amplify a PCR product of the predicted size, this will indicate that you have successfully generated functional primers for the gene of interest. The next step to discovering a new sequence will be to isolate the PCR product from the gel and have it sequenced. If you can do this, you will be making a contribution to our current knowledge of genetics!

7

Using PCR to
Find New Gene Sequences

INTRODUCTION

In a previous lab activity, instructions were given for designing novel PCR primers. In this lab, you will utilize a pair of novel primers to amplify a previously unidentified DNA region. Instructions are given for isolating DNA from an animal species (an insect) and a plant species. If you are successful in amplifying a good PCR product from a gene that has not previously been sequenced in a particular organism, it may be possible to have the DNA sequenced, thus adding to the knowledge base for the organism.

PROCEDURE

Although instructions for DNA isolation are given here for both a plant and an animal, you will only work with one organism. The steps following the DNA isolation process will be identical for all DNA samples.

DNA Isolation for Insects

The following procedure could be used with various types of insects; however, it was originally designed for the fruit fly, *Drosophila*. For this reason, you are most likely to have success if you work with a small fly, gnat, or similar insect for this portion of the experiment (if the insect you use is significantly larger than *Drosophila*, you should scale up the reactions to adjust for the size difference). Alternatively, to obtain higher-quality DNA, your instructor may choose to have you utilize a commercial DNA extraction kit. One that has given very good results for insect DNA is a kit sold by Cartagen called Genomic DNA Extraction Kit: Arthropods (catalogue #:20810-050).

Step 1: Each student should isolate DNA for one or more insects as instructed. Label one microcentrifuge tube for each insect with a code number as instructed. Place one insect in each labeled tube. Note the code number of the insect(s) you are working with for future reference.

Step 2: Add 50 µl of insect homogenization buffer to each tube, and grind the insect with a plastic pestle. Grind the insect as completely as possible. Use care when grinding and when removing the pestle to leave as much homogenate in the bottom of the tube as possible.

Step 3: Incubate the tubes for 25 minutes at approximately 30°C. This can be accomplished by placing the tubes in a floating tube "rack" in a 30°C water bath or in a heat block set for 30°C.

Step 4: Incubate the tubes for 1 to 2 minutes at ca. 95°C using a similar procedure. Use care when removing the tubes from the hot water.

Step 5: If a microcentrifuge is available, briefly spin the homogenized insects so that remaining tissues form a pellet at the bottom of the tube. Keep this solution and any dilutions made from it on ice. The supernatant from this crude homogenate can be used directly as a source of DNA in your PCR reaction.

DNA Isolation for Plants

The following procedure was originally designed for use with the model plant *Arabidopsis*. For the best chance of obtaining good results, you should use plant material derived from young seedlings, since young plant tissues have fewer inhibitory substances in them. Alternatively, to obtain higher-quality DNA, your instructor may choose to have you utilize a commercial DNA extraction kit. One that has given very good results for plant DNA is a kit sold by Mo Bio Laboratories, Inc., called UltraClean Plant DNA Isolation Kit (catalogue #:13000-50).

Step 1: Each student should use a piece of plant tissue that is roughly equivalent in weight to about 10-20 mg. Depending on the size of the seedlings you are working with, this may represent a portion of a leaf, one or more entire leaves, or the entire above-ground portion of a small seedling.

Step 2: Place the plant material in a 1.5 mL microcentrifuge tube. Label the tube with a code number as instructed. Note the code number of the plant(s) you are working with for future reference.

Step 3: Use a clean plastic pestle to completely grind the tissue for about 30 seconds.

Step 4: Add 400 μl of plant homogenization buffer to the plant material and grind briefly again to thoroughly mix the plant material with the solution.

Step 5: Close the lid of the microfuge tube and vortex it (or vigorously shake it) for about 5 seconds.

Step 6: Heat the sample for 5 minutes in a water bath or a heat block set for 95°C (use care to prevent exposing your skin to the extreme heat).

Step 7: Place the tube in a balanced microcentrifuge and spin it for 2 minutes.

Step 8: Without disturbing the pellet, transfer 350 μl of the supernatant into a fresh tube (discard the old tube with remaining plant material).

Step 9: Add 400 μl of isopropanol to the tube containing the supernatant.

Step 10: Mix the contents of the tube by inverting it several times, and then leave it at room temperature for 3 minutes.

Step 11: Place the tube in a balanced microcentrifuge with the hinge pointing inward (i.e., toward the inside of the rotor) and spin the tube for 5 minutes. Orienting the microcentrofuge with the hinge pointed in will ensure that the DNA pellet will form on the side of the tube opposite of the hinge (this is helpful to know when you remove the supernatant).

Step 12: Carefully pour off the supernatant from the tube, and then carefully remove any remaining supernatant using a micropipette. Remember that the DNA will be at the bottom of the tube on the side opposite from the hinge – be careful not to disturb it.

Step 13: Air dry the pellet for about 10 minutes to allow any remaining isopropanol to be completely evaporated.

Step 14: Add 100 µl of TE buffer to the tube and dissolve the DNA pellet by repeatedly aspirating the solution up and down. Ensure that any visible pellet at the bottom of the tube becomes completely resuspended.

Step 15: Keep this solution and any dilutions made from it on ice. The supernatant from this crude homogenate can be used directly as a source of DNA in your PCR reaction.

PCR Reactions

From this point on, both types of DNA can be treated in the same manner (i.e., using the same volumes, reactions times, etc.). A standard process for setting up and running PCR reactions is given below, but first, three factors — DNA concentration, annealing temperature, and using control primers — must be considered.

DNA Concentration

Your PCR reactions should ideally contain about 1 µg or less of template DNA (however, reactions often work fine with higher concentrations). If a high-quality spectrophotometer is available, you may directly quantitate the concentration of a dilution of your DNA sample (your instructor will show you how to use the spectrophotometer). You can then dilute your DNA to approximately 0.5 µg per µl (and use 2 µl in your PCR reactions).

Remember that the concentration you obtain from the spectrophotometer analysis will indicate the concentration of your dilute solution, so you will need to multiply this number by your dilution factor to obtain the concentration of your full-strength DNA solution.

Even if you know the DNA concentration of your sample, it is usually wise to run each PCR condition (each pair of primers) in several concentrations, including a negative control with no DNA. This way, at least one concentration should be in the proper range for good PCR amplification (it is difficult to obtain reliable spectrophotometric measurements of DNA). After your DNA is diluted to 0.5 µg/µl, make six PCR reactions for each DNA sample with 0 (negative control), 0.5, 1.0, 1.5, 2.0, and 3.0 µl of DNA solution per reaction. You will add 3.0, 2.5, 2.0, 1.5, 1.0, and 0 µl of distilled water respectively to each tube to bring the total of each tube to 3.0 µl of DNA/water solution.

If you cannot quantitate your DNA sample, make a 10-fold dilution and a 100-fold dilution of your original sample, and make the following eight conditions (adding appropriate volumes of distilled water to bring your sample up to 3 µl): 0 µl (negative control), 3.0 µl of the 100-fold dilution, 3.0 µl of the 10-fold dilution, and 0.5, 1.0, 1.5, 2.0, and 3.0 µl of the original sample.

In the next section, you will consider adding several annealing temperatures for each DNA concentration. Since the number of possible reactions could become quite large, your instructor may decide to have you reduce the number of DNA concentrations you will test.

Annealing Temperature

You will base the annealing temperature for your PCR reaction on the temperature that was specified in the primer design program when the primer sequences were generated. The program that generated these sequences used known rules to find sequences that would anneal specifically at the temperature specified. However, the calculations yield only theoretical optimal primers for the specified temperatures. Often, the true optimum annealing temperature for the selected primers is 1° or 2° above or below the predicted optimum temperature. For this reason, for each PCR condition (i.e., for each pair of primers and for each DNA concentration), you should test several annealing temperatures. Some PCR thermal cyclers have a gradient block, which will allow an investigator to test several annealing temperatures at the same time (the block can provide a range of temperatures over its surface). However, if your PCR thermal cycler does not have this feature, you will have to run a separate PCR reaction for each annealing temperature you use.

Verify with your instructor the number of annealing temperatures you should use. Now, take a minute to make sure that you know how many PCR tubes you will be preparing to test your primers (consider both DNA concentrations and annealing temperatures). In the space below, use the headings shown to create a table with three columns. This table will show code designations for each PCR tube (which you will use when you label your PCR tubes), details regarding the DNA concentration for each tube, and the annealing temperatures that will be tested in each case. Since it may be necessary to test a fairly large number of conditions, your instructor may have you plan and test these conditions as one group or as part of several large student groups.

PCR Tube Code Designation	Notes on DNA Concentration	Annealing Temperature

Using Control Primers

If your PCR reactions produce appropriately sized bands in an electrophoresis gel, you will know that your DNA isolation technique was good and that your primers were correctly designed. However, if no bands are apparent, this alone will not indicate whether your DNA was of poor quality or, alternatively, whether your primers were not correctly designed (or possibly whether the annealing temperatures were all too high or too low). To identify the source of error if your PCR reaction fails to generate a product, you should use a control primer pair known to amplify a PCR product correctly for the DNA you are using.

Since the 18S ribosomal gene is highly conserved, primers for this gene can be generated that should amplify products in virtually all eukaryotic organisms. As a control for this experiment, you should run duplicate samples of all of the DNA concentrations listed in the table earlier, using a pair of 18S ribosomal gene primers as a control. A single optimum annealing temperature can be determined for these primers; therefore, you should only need to run one annealing temperature condition for this primer pair. This primer pair should amplify a product in one to several of your PCR tubes (based on correct DNA concentration) if your DNA sample is sufficiently pure and undegraded. If you get good bands with this control primer pair, but not with your other primer pair, then it is probable that either your experimental primers are not complementary with the sequence in your template DNA, or the optimum annealing temperature is actually lower or higher than the temperatures you tried with your experimental primers.

Preparing Standard PCR Reactions

Step 1: Add appropriate amounts of template DNA and distilled water (as needed) to labeled tubes, as described earlier. Each tube should contain a total of 3 µl of DNA/water solution. Keep your PCR tubes on ice during this process, until you place them in the DNA thermal cycler (for longer storage, they can be kept refrigerated until thermal cycling).

Step 2: Add 22 µl of PCR "master mix" to each of the DNA samples. There will be two master mixes, one for your experimental primers and one for the 18S ribosomal gene controls. Be sure to add the correct mix to each tube.

The PCR master mix contains the following components: A buffer, a thermostable DNA polymerase, KCl and $MgCl_2$ (to serve as cofactors for the polymerase), a dNTP mix, and each of the primers specific for the particular PCR reaction.

Step 3: If your thermal cycler requires oil (your instructor will advise you of this), add 1 to 2 drops of mineral oil to the side of the PCR tube. It will form a layer over the aqueous PCR mixture.

If you added oil, "burst"-spin the PCR tube in a microcentrifuge to completely separate the aqueous PCR mix from the oil. *Caution:* Place PCR tubes within the "larger" microcentrifuge tubes *with the microcentrifuge tube caps cut off* within the microcentrifuge; otherwise, the PCR tubes will not remain in the rotor. Spin the tubes for several seconds.

Step 4: Your instructor will set up the thermal cycler to conduct PCR.

Electrophoresis of PCR Samples

Note: This will be done on a different day from the preceding procedures.

Step 1: Add 5 µl of a *6X* loading dye to the aqueous portion of each PCR sample. Burst-spin as done in Step 3 (above) to separate the oil from the dyed aqueous solution. The dye serves both to make your sample more viscous and colored (to facilitate gel loading) and to provide a visible reference to determine how long to conduct electrophoresis.

Step 2: Set your micropipette for the volume indicated by your instructor (this volume differs with gel type). Carefully submerge the tip of your micropipette below the oil layer (if present), and slowly depress the plunger, expelling small air bubbles into the dyed aqueous solution. Then *slowly* release the plunger, aspirating the measured amount of the dyed PCR solution. Raise the tip above the oil layer and move the tip around the upper wall of the tube to remove excess oil adhering to the outside of the tip.

Step 3: Add your sample to one well in the gel (note which well[s] you load), by submerging the tip below the buffer solution and directly into the opening of a well. Do not expel the solution until you are sure that your tip is inserted into a well, and do not insert your tip too far into the well (be careful to avoid puncturing the bottom of the gel with your pipette tip). When you expel the sample into the well, you should be able to see the colored solution settling into the well and filling it. One student should also load one lane on each gel with a molecular marker for size determination.

Step 4: After the gel is loaded, your instructor will set the voltage. The gel should be allowed to run until the dye front is close to the opposite end of the gel.

Analysis of PCR Products

Step 1: View the gel(s) on an ultraviolet light box, or use a gel documentation system to photograph the gel (wear gloves when handling the gel — it contains a potent mutagen, ethidium bromide). If you view the gel directly (rather than viewing an image of the gel), *wear appropriate plastic eye protection*. Since prolonged exposure to UV light may also harm the skin, students may wish to wear a plastic face shield and appropriate clothing to cover exposed skin.

If your PCR experiment is successful and some of your reactions produce clear, obvious bands in the electrophoresis gel, you should estimate the size of the PCR products. To do this, take measurements of how far each band migrated from the well. Measure the distance from the well to the center of the band on the gel that represents your PCR product. Also measure the distances between the well and each of the bands in the lane that contains the reference DNA. Your instructor will tell you the sizes of the DNA fragments represented by each band in the reference DNA lane.

Step 2: To estimate the size of the DNA you isolated, compare the distance traveled by the band representing the PCR product to that of each of the reference bands. Do this by plotting a standard curve for the reference DNA fragments on semilog graph paper and comparing the measurement for your PCR product to these. Plot the sizes of the reference fragments (in base pairs) on the log scale axis and the distances migrated by the bands on the standard scale axis. You should be able to draw a straight line passing through (or nearby) each of the points. Now use the standard curve to estimate the size of the PCR product you generated.

Q1 How large was the PCR product you generated? Hand in a copy of your standard curve showing distance migrated by your PCR product and its estimated size on the graph paper on page 62.

If this product is of the expected size, you may have isolated a new gene sequence! The next step, if time and resources permit, may be to have your PCR product sequenced, thus contributing to the knowledge base for this organism.

NOTES

NOTES

Mapping a Molecular Marker
in *Drosophila*

8

INTRODUCTION

In this lab activity, you will study the inheritance of a molecular marker in *Drosophila melanogaster* Specifically, you will conduct an experiment to demonstrate that the marker is linked to a particular gene for a visible characteristic in *Drosophila*. This experiment demonstrates one of the strategies that is often used in the early stages of genome characterization in humans and other organisms. In this experiment, you will study the inheritance of the marker and that of two genes associated with physical mutations, *black* body and *Bar* eye. Your goal will be to demonstrate that the marker is linked to the *black* body gene and to estimate from the data how closely the two loci are linked.

STRATEGIES FOR GENOME CHARACTERIZATION

Information to characterize an organism's genome can be obtained in various ways. One of the most obvious is by sequencing some or all of the organism's DNA. However, this step is generally preceded by other steps that provide more general information about genome organization. For example, since the function of most DNA from a given organism is often unknown, it is generally necessary to begin by locating specific genes.

One way to locate specific genes is to use molecular markers to identify these genes. Various kinds of molecular markers can be utilized in different ways. Some are very specific, such as nucleic acid probes. These probes recognize and bind with unique DNA sequences. However, to generate DNA probes, some sequence information for the gene in question is required. This is possible if, for example, the gene of interest has already been identified and sequenced in a closely related species. In this case, it is likely that the gene sequences would be similar enough that the sequence data from one species could be used to design a probe to locate the gene in the other organism.

However, it is also possible to find genes for which no sequence information is yet available. This can be done by using markers that are based on random genetic polymorphisms (polymorphisms are variations in length or other attributes). Such an approach was used to locate several human genes associated with genetic disorders during the early stages of the Human Genome Project. (The gene associated with cystic fibrosis was discovered this way; the story of the discovery of this gene can be found at http://www.hhmi.org/genetictrail/a100.html.) Some polymorphisms used for markers are based on the presence or absence of particular restriction enzyme sites. These markers provide characteristic "genetic fingerprints" for individuals because restriction enzymes cut the DNA into different-sized fragments.

Another type of molecular marker is based on polymorphisms that are due to actual variation in the length of a particular region of DNA (not just the presence or absence of a restriction site). You will be using such a marker — a *microsatellite sequence* — in this lab activity. In microsatellite sequences, the length variations are due to the presence of small nucleotide sequences that are repeated several to many times. The length variations can be detected by using a polymerase chain reaction (PCR) to amplify the microsatellite sequence (if you are unfamiliar with PCR, please review the description presented in Lab Activity 6, entitled "Designing PCR Primers"). The PCR primers used to amplify these sequences flank the regions where the repeats occur. If fewer repeats are present, the resulting PCR product is shorter; if more repeats are present, it is longer.

Although the significance of most microsatellite variations is unknown, some that occur in humans are associated with genetic diseases, such as Huntington's disease. Many other microsatellites in humans and other organisms appear to have no effect on health or viability. However, they are of interest to geneticists because they can be used as molecular markers. For example, when geneticists search for the location of genes associated with diseases, they begin by screening members of families that carry the disease, looking for patterns associated with large numbers of molecular markers. What they hope to find is a pattern that will indicate that one or more of the markers is linked to the disease-causing form of the gene.

To illustrate, consider the example of a family that carries a hypothetical dominant condition (see Figure 8.1). The condition is exhibited by the father and three out of seven siblings. In this case, the father carries a molecular marker (the 8 kb fragment in Figure 8.1) that is also present in the three affected siblings. Significantly, the mother lacks this marker, and so do the unaffected siblings. In this example, it seems quite likely that the marker is physically associated with the gene that causes the disease. Further confirmation could be obtained by finding a similar pattern among families of any affected brothers or sisters of the affected father in Figure 8.1. The situation illustrated in Figure 8.1 suggests that the marker is very closely linked to the gene in question (because all affected individuals carry the marker). In most cases, a linked marker would occur more frequently with a given phenotype than apart from it, but it would not always be associated with it because of genetic recombination.

Most genetic analyses of this type would be more complex than the situation described in Figure 8.1. For one thing, it is usually necessary to screen large numbers of markers before finding one that shows a pattern suggesting linkage. However, despite the complexity, this approach often pays off, yielding a marker that is physically linked to a gene of interest. If the marker has been well characterized, its location on a particular chromosome may already be known. In this case, the location of the gene of interest can be narrowed to a single chromosome. Furthermore, whether the location of the marker is known or not, the marker itself can be used as a nucleotide probe. Such a probe could then be used to screen a genomic DNA library, which could lead to the isolation of the actual gene of interest itself.

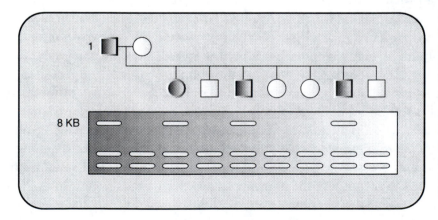

Figure 8.1 – The upper portion of the figure shows a pedigree for a family carrying a gene for a hypothetical dominant condition. As indicated by the shaded symbols, the father and three of his children are affected by the condition. The bottom half of the figure shows the electrophoresis patterns for a molecular marker (the 8 kb fragment) and other nonspecific bands for each individual in the pedigree.

BACKGROUND INFORMATION FOR THE LAB PROCEDURE

Since it is somewhat difficult to conduct genetic experiments with human subjects, in this lab, you will work with a molecular marker present in *Drosophila*. Furthermore, for the purpose of this lab activity, it is not feasible to conduct many experiments using a large number of randomly selected molecular markers and different strains of fruit flies. Therefore, you will be given a specific cross to analyze, and you will know ahead of time that the molecular marker is linked to the *black* body gene. However, it is still important that you understand how this type of analysis works to comprehend how it shows that the marker and the *black* gene are linked. In addition, you will carry the analysis one step further, by using the data you obtain to determine how many map units separate the marker and the *black* body gene.

Your class will be conducting the molecular analysis with the parents and offspring from a test cross involving the two strains carrying different versions of the molecular marker (see the details given later). The test cross will utilize F1 female flies that are heterozygous for the genes associated with *black* and *Bar* (and also heterozygous for the molecular marker those strains carry — discussed later) and male flies that are homozygous for the *black* mutation (and also homozygous for one form of the molecular marker).

Your instructor may choose to give you the parental flies (*black* body and *Bar* eye) and have you conduct all the steps needed to prepare the cross. Alternatively, you may be given the F1 flies and the black flies and instructed to test-cross them, or you may simply be given the test-cross progeny to analyze. The first and second alternatives require more time (at least 4 to 5 weeks for the first alternative) and a great amount of care to obtain meaningful results. If your instructor chooses to have you conduct all or some of the steps necessary to obtain the resulting progeny for the cross, you should consult a separate exercise in this manual (Lab Activity 3, entitled "Introduction to *Drosophila* and Conducting Crosses") to learn appropriate techniques for mating fruit flies. Since obtaining meaningful results will require the collection of large amounts of data, this experiment should be conducted as a joint effort by an entire lab section or genetics class.

The Strains and the Genotypes

The two strains of flies that serve as original parental strains in this cross are each homozygous for a physically observable phenotype as well as a molecular marker (distinguishable by PCR and electrophoresis — discussed later).

The *black* Parents

The *black* parental strain is homozygous for the recessive allele that produces a black body color. Since this is a recessive phenotype, the symbol for the black allele is "b," and black fruit flies have the genotype "bb." Instead of using a capital *B* to indicate the dominant form of this gene (which produces normal, brownish, body color), *Drosophila* geneticists use the symbol + to indicate all normal (wild-type) alleles. Using this system, the three possible genotypes for this character are ++, +b (which both exhibit the dominant, wild-type phenotype), and bb (which exhibits the recessive, black phenotype). Using this system, you can tell just by looking at the three possible genotypes that the phenotype being studied is a recessive one (because a lowercase *b* is used as the symbol for black).

The *black* parental strain is also homozygous for the form of the molecular marker that produces the "short" (about 132 base pairs) PCR product. We will symbolize the genotype for this trait in the black flies as "SS" (flies that are homozygous for the longer form of the marker will be symbolized as "LL," and heterozygotes will be symbolized as "LS"). You can't tell by looking at one of these black parental flies that it has the genotype SS. The only way you can tell is by extracting its DNA, amplifying it with PCR, and then running the PCR product on an electrophoresis gel. DNA from a fly with the genotype SS will yield a PCR product that will produce only one band on a gel. DNA from an LL fly will yield a product that will produce one band also, but the band will be closer to the top of the gel (because it is a longer PCR product), and DNA from an LS fly will yield two bands.

The *Bar* Eye Parents

The allele associated with *Bar* eye is an X-linked "semidominant" allele. Because it's on the X chromosome, only females can have two copies of the mutant allele (BB). Such females have extremely narrow eyes that are about one-third the width of the eyes of wild-type flies. Females with the genotype +B have slightly wider eyes, but they are still narrower than those of wild-type flies (++). Males can only have two genotypes for this gene, since it is an X-linked gene. The two genotypes can be symbolized as +Y and BY (where "Y" indicates the Y chromosome); such flies have normal eyes and somewhat narrowed eyes, respectively.

The *Bar* eye parental flies are homozygous for the "long" form (about 165 base pairs) of the molecular marker, having the genotype LL. Remember that "S" and "L" are alternative forms of the same marker. Also remember that the microsatellite marker is at a different locus from either *Bar* or *black* (although it is linked to *black*).

The Test Cross

This cross is conducted by using virgin females that are heterozygous for all three loci (F1 flies; genotype +B, +b, LS) and male flies from the *black* parental strain (+Y, bb, SS). Only females are used as the heterozygous parent because male *Drosophila* do not exhibit genetic recombination. To estimate the number of map units that separate the *black* locus from the molecular marker, it is necessary to determine the percentage recombination that occurs between these two loci.

Expected Results: Testing the Null Hypothesis

To analyze the results of this cross, you will need to assess each fly from the test cross progeny for two characteristics. These characteristics include the visible one, which is body color, and the molecular characteristic, which is number, and types of PCR products for the molecular marker (LL, SS, or LS). This means that you will be observing each fly and recording its body color, then you will be extracting its DNA, preparing a PCR reaction, and characterizing its "molecular genotype" using gel electrophoresis. The results that you should expect in this case will be observations that support the hypothesis that the molecular marker is, in fact, linked to the gene for *black* body (you will also be quantifying the apparent amount of genetic recombination).

The simplest way to infer that linkage exists between two genetic loci is to disprove the null hypothesis, which, in this case, is that the two loci are not linked. If you can clearly identify what results the null hypothesis would predict, it becomes easy to identify results that would disprove the null hypothesis (implying that the two genes are linked). To help you visualize this relationship, consider Figure 8.2. This figure illustrates one possible interpretation of the test cross you will analyze. It is based on the null hypothesis that the gene for *black* body and the locus for the molecular marker are unlinked (realize that this null hypothesis is, in fact, incorrect).

Now, to consider the implications of the null hypothesis and to understand the kinds of data you will expect to observe if the null hypothesis is incorrect (as expected), answer the following questions:

Q1 Consider the first category of test-cross offspring shown in Figure 8.2 (+b, LS).

   A.   What would be the physical phenotype of these flies?

   B.   If PCR was conducted with the DNA of one of these flies using the primers for the molecular marker, what would be the appearance of the bands on an electrophoresis gel with the PCR products?

Q2 If Figure 8.2 was accurate (i.e., if the null hypothesis was correct), (a) what proportion of the test-cross progeny would be black flies that are heterozygous for the molecular marker? (b) What proportion would be flies with normal body color, which are homozygous for one form of the molecular marker?

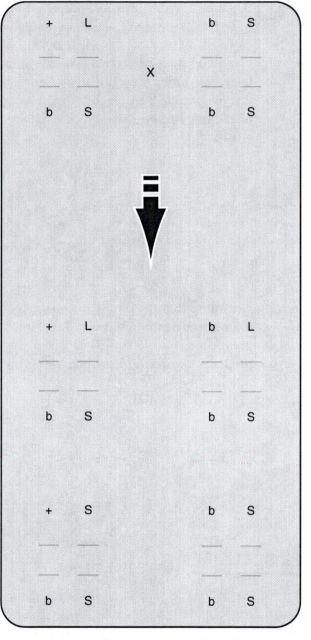

Figure 8.2 – An illustration of results of a test cross for the cross involving the black gene and a molecular marker. The figure illustrates how the alleles would assort if the null hypothesis were correct (i.e., if the gene for *black* and the locus of the molecular marker were unlinked). Each horizontal line represents a single chromosome.

Q3 If the null hypothesis is incorrect (as expected), the same four genotypes should be present in test-cross progeny, as shown in Figure 8.2; however, the proportion of each type of progeny will differ (and the arrangement of the genes on the chromosomes will be different). Describe in general terms how the proportion of flies in each genotypic category will be different assuming the null hypothesis is incorrect (describe how the number of flies *in each category* would be expected to change). *Note:* It is important to remember that the parents of the heterozygous female flies in the test cross had the following genotypes: bb, SS, and ++, LL.

Q4 Assuming that the data you collect do, in fact, indicate that the *black* gene and the locus for the molecular marker are linked, how can you estimate the number of map units that separate these two loci? Write out a formula (using expected categories from the test-cross progeny) indicating how you will estimate this value.

PROCEDURE

Either you will be given flies representing the progeny of the test cross described earlier, or your instructor will have you generate these progeny via the appropriate crosses. Together with the rest of your class, you should analyze as many of the test-cross progeny as possible (it should be possible for each student to process several flies, if necessary). Each fly should be given a code number that will identify it in your records and will also be used to designate its DNA and the resulting PCR product. The instructions for Steps 1 through 3 are given here for individual flies.

Step 1: Observe the anesthetized fly using a dissecting microscope. Record a code number for this fly and record its body color.

Step 2: Grinding individual flies in a buffer solution

a. Label a microcentrifuge tube with the appropriate code number representing the fly, and place the anesthetized fly in the tube.

b. Add 50 µl of homogenization buffer to the tube, and grind the fly with a plastic pestle as completely as possible. Use care when grinding and when removing the pestle to leave as much fly homogenate in the bottom of the tube as possible.

c. Incubate the tube for 25 minutes at approximately 30°C. This can be accomplished by placing the tube in a floating tube "rack" in a 30°C water bath.

d. Incubate the tube for 1 to 2 minutes at about 95°C using a similar procedure. Use care when removing the tubes from the hot water.

e. If a microcentrifuge is available, briefly spin the homogenized fly so that remaining tissues form a pellet at the bottom of the tube. Keep this solution on ice during subsequent steps.

Step 3: Preparing and amplifying PCR reactions for each fly

a. Label a PCR tube with the appropriate code and transfer 2 µl of Fly DNA solution into it.

b. Add 23 µl of PCR "master mix." The master mix contains all the components needed for a PCR reaction, except the Fly DNA (added in Step a). The final PCR reaction contains the following: 10 mM Tris-HCl (pH 8.3), 0.75 units of *Taq* polymerase, 50 mM KCl, 1.5 mM $MgCl_2$, 0.2 mM dNTP mix, and 50 pmoles of each primer.

c. If your thermal cycler requires oil (your instructor will advise you of this), add 1 to 2 drops of mineral oil to the side of the PCR tube. It will form a layer over the aqueous PCR mixture.

d. If you added oil, burst-spin the PCR tube in a microcentrifuge to separate the aqueous PCR mix completely from the oil. *Caution:* Place PCR tubes within the "larger" microcentrifuge tubes *with the microcentrifuge tube caps cut off* within the microcentrifuge; otherwise, the PCR tubes will not remain in the rotor. Spin the tubes for several seconds.

e. Your instructor will set up the thermal cycler to conduct PCR.

Step 4: Conducting electrophoresis of PCR samples (this will be done on a different day from the preceding procedures)

a. Add 5 µl of loading dye to the aqueous portion of each PCR sample. Burst-spin as done in Step 3 to separate the oil from the dyed aqueous solution if necessary.

b. Set your micropipette for the volume indicated by your instructor. Carefully submerge the tip of your micropipette below the oil layer (if present) and slowly depress the plunger, expelling small air bubbles into the dyed aqueous solution. Then *slowly* release the plunger, aspirating the measured amount of the dyed PCR solution. Raise the tip above the oil layer and move the tip around the upper wall of the tube to remove excess oil adhering to the outside of the tip.

c. Add your sample to one well in the gel (note which well[s] you load), by submerging the tip below the buffer solution and directly into the opening of a well. Do not expel the solution until you are sure that your tip is inserted into a well, and do not insert your tip too far into the well (be careful to avoid puncturing the bottom of the gel with your pipette tip). When you expel the sample into the well, you should be able to see the colored solution settling into the well and filling it. One student should also load one lane on each gel with a molecular marker for size determination.

d. After the gel is loaded, your instructor will set the voltage.

e. The gel should be allowed to run until the dye front is close to the opposite end of the gel (unless a somewhat long gel is used). The gel should be viewed immediately following electrophoresis.

Step 5: Characterizing the bands present on gels — View the gel(s) on an ultraviolet light box or use a gel documentation system to photograph the gel (wear gloves when handling the gel — it contains a potent mutagen, ethidium bromide). If you view the gel directly (rather than viewing an image of the gel), *wear appropriate plastic eye protection.* Since prolonged exposure to UV light may also harm the skin, students may wish to wear a plastic face shield and appropriate clothing to cover exposed skin.

The pattern in each sample lane should represent either a "long/long" homozygote (one band closer to the top of the gel), a "long/short" heterozygote (two bands), or a "short/short" homozygote (one band closer to the bottom of the gel). To verify that appropriately sized bands were produced by PCR, you should compare the sample bands to the DNA in the molecular marker lane. Depending on the efficiency of your DNA isolation technique, there may be one or more lighter bands in addition to the band(s) of expected size; however, in most cases, one or two bright bands should be obvious.

To accomplish the analysis of the data from this experiment, you must determine and tabulate the body color and the molecular genotype (LL, LS, or SS) of each fly. You should do this in a large group with the assistance of your instructor. Each student will then analyze the entire data set and answer the questions that accompany this lab activity.

Q5 **Indicate the number of flies in each of the four genotypic categories among the test-cross progeny (i.e., +b, LS; +b, SS; bb, LS; bb, SS).**

Q6 **How many map units appear to separate the locus for the *black* gene from the locus for the molecular marker? (Show the calculations you used to arrive at this number.)**

LAB ACTIVITY

Yeast as a Model System:
Conditional Mutants

9

INTRODUCTION

The single-celled, eukaryotic organism *Saccharomyces cerevisiae*, commonly known as baker's yeast, is an enormously important model organism in the study of genetics. It exhibits many of the complex characteristics of other eukaryotes, but it is extremely easy to grow and manipulate. Because it is a single-celled organism, it can be grown and analyzed using simple techniques similar to those used with bacteria. In recent years, geneticists have used yeast to study the details of intricate phenomena, such as the control of the cell cycle and chromosome composition and organization. In 1996, *Saccharomyces* became the first eukaryote for which an entire genomic sequence was available.

In this laboratory exercise, you will work with yeast cultures to investigate the life cycle of this organism and to study several conditional mutations.

THE YEAST LIFE CYCLE

Yeast may exist as either haploid or diploid cells (see Figure 9.1) that can be grown in liquid medium or on agar-solidified medium. Reproduction can occur either by asexual means (budding) during either stage or through fusion of two haploid cells, followed by subsequent meiosis. There are two mating strains, a and α, which must be present together in order for fusion to occur. When the two mating types are cultured together, diffusable pheromones produced by both types induce changes in the cells preparing them for fusion. Part of the change is a shape change that causes the otherwise spherical yeast cells to become pear shaped. Because of their resemblance to a mythical "animal" featured in a comic strip called "Lil' Abner" (now long out of publication), these cells are called "shmoos" (see www.lil-abner.com/ for a cultural history lesson). After the haploid cells fuse, they form a peanut-shaped zygote. The resulting diploid cell can reproduce indefinitely by mitosis, or it may be induced to undergo meiosis, forming haploid cells (spores) again. Meiosis can be induced by growing the diploid cells on a nitrogen-deficient growth medium.

Q1 Consult Figure 9.1. Following meiosis, which mating type(s) would you expect to be present in the spores produced? If more than one mating type would be present, what ratio of the two mating types would you expect to exist?

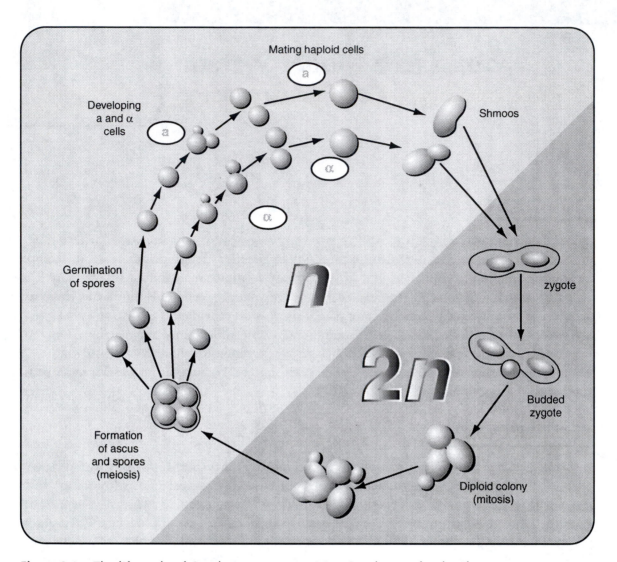

Figure 9.1 – The life cycle of *Saccharomyces cerevisiae*. See the text for details.

Conditional Mutations

Conditional mutations are an interesting and useful category of mutations in which the mutant phenotypes are only expressed under certain conditions.

This type of mutation is useful to geneticists for several reasons. For one thing, such mutations can be turned "on" and "off" under different conditions. This allows a geneticist to compare things such as the biochemical composition of the organism under both conditions. Also, many conditional mutations are lethal under some conditions. However, since they are conditional mutants, they are viable under other conditions, allowing researchers to conduct long-term studies on these lethal mutants. Two common types of conditional mutations are temperature-sensitive mutations and nutritional mutations.

Various temperature-sensitive strains, exhibiting different types of phenotypes are available for most model organisms. Some may produce an alteration in a visible phenotype under different temperatures or even lead to an altered behavior. Among the most common types of temperature-sensitive mutations are lethal or sublethal mutations. In strains carrying such mutations, biological processes occur normally at certain temperatures (called *permissive* temperatures), but the organism either dies or grows more slowly at other temperatures (*restrictive* temperatures). Such mutations may prevent (or induce) expression of a particular gene at the restrictive temperature, or they may result in the production of a protein (gene product) that is unstable, or otherwise functionally different, at the restrictive temperature.

Many nutritional mutations also exist for a wide variety of organisms. In simple organisms, such as bacteria and fungi, strains with these mutations are designated as *auxotrophic*, as opposed to *prototrophic*. The wild-type (prototrophic) form of such organisms can usually survive when grown on a simple nutrient medium (a "minimal medium") containing minerals and a few organic compounds, but lacking many complex compounds, such as certain amino acids or vitamins. Auxotrophic strains are able to grow normally on a "complete medium" that contains all the components in the minimal medium plus certain complex organic substances. However, such strains can only grow on minimal medium if it is supplemented with the one substance they are unable to produce.

The Nobel Prize-winning work of Beadle and Tatum in the early 1940s that used the fungus *Neurospora* is a classic example of a research strategy using strains of auxotrophic mutants. Virtually all general genetics textbooks describe their research. If you are not already familiar with this work, you should take time to review it now. Part of the work you will do in this lab activity will be similar to some aspects of the studies conducted by Beadle and Tatum.

PROCEDURE

Observing the Life Cycle of *Saccharomyces*

Your instructor will give you two small, coded samples of yeast growing on agar-solidified medium. One sample will contain only haploid cells from one of the two mating strains. The other sample will contain a mixture of haploid cells from both mating strains, as well as other stages of the life cycle. The sample with the mixture of cells will have been mixed about 3 hours prior. You should prepare wet mount slides for each of these samples and compare them using a compound microscope. Use several drops of water and a small amount of yeast to make the wet mount slides.

 Q2 Sketch (a) a haploid cell that has not been induced to begin the mating process, (b) a shmoo, and (c) a peanut-shaped zygote. Which coded sample contained the two different mating strains?

Working with Conditional Mutants

Temperature-Sensitive Mutants

Your instructor will give you two or more coded Petri plates containing samples of yeast with unknown phenotypes and a control plate containing a wild-type strain. One or more of the strains in the coded plates will be temperature-sensitive mutants. For wild-type yeast, the optimum growth temperature is about 30°C, but good growth can occur over a range of temperatures, from below 25°C to above 37°C. A permissive temperature for temperature-sensitive yeast mutants is generally about 25°C (room temperature is fine), and the typical restrictive temperature is 37°C or above. You should attempt to grow each strain you are assigned and the control strain at a permissive and a restrictive temperature.

Step 1: For each strain you will be testing, label two Petri plates containing fresh medium with the code designation for that strain, and the temperature at which you will be growing it. Two or more samples of yeast can also be grown in a single Petri plate by sub-dividing it into several sections (by marking the bottom of the plate).

Step 2: Transfer some of the yeast cells onto the fresh plate using a sterile toothpick or a sterile probe. Use sterile technique to avoid introducing foreign contamination. If two or more samples are grown together in a single Petri plate, use care to keep them physically separated.

Step 3: For each strain transferred to fresh medium, streak the transferred cells over the surface of the medium by gently rubbing the surface of the medium, dragging the toothpick or probe through the transferred cells and over previously untouched parts of the medium. Make a series of streaks over the surface using a zig-zag motion to spread the cells.

Step 4: Incubate your cultures at appropriate temperatures, and make observations on them over the course of the next several days. Depending on the temperature at which the yeast are incubated, viable cells should grow enough to see differences by one to several days. As you assess the amount of growth that has occurred for an "unknown" strain at a given temperature, you should compare its growth with that of the control culture at the same temperature and also with that of the unknown at the permissive temperature (see the following note).

Note: It is possible that a small number of cells may have reverted to the wild-type phenotype (i.e., a reverse mutation may have occurred) in some of the cultures that originally contained all temperature-sensitive mutant cells. This is often seen in yeast cultures because reproduction occurs so rapidly. Although reverse mutations are rare, a small number of revertants are likely to exist in cultures containing millions of cells. Therefore, the presence of a small number of growing yeast colonies should not be taken as evidence that the original strain was not a temperature-sensitive strain. Growth of experimental cultures should always be compared to that of control cultures.

Q3   **A. Which coded culture(s) contained temperature-sensitive mutants?**

B. Did the mutant strain(s) appear to be temperature-sensitive lethals or sublethals? Explain.

Auxotrophic Mutants

Your instructor will give you one to several coded Petri plates with complete medium, containing samples of haploid, auxotrophic yeast. You will also receive a control plate containing a wild-type strain. The coded samples are all "unknown" auxotrophic strains. Each of these strains lacks the ability to synthesize some necessary organic compound normally produced by the wild type. You will be conducting an experiment to determine which of several organic compounds each strain requires. For each unknown sample assigned to you, you will prepare several culture vessels, each containing minimal medium plus one of the organic compounds supplied by your instructor. You will also grow each strain on complete medium and wild-type cells on each type of medium as controls.

The procedure you will be conducting is similar to a portion of the research conducted by Beadle and Tatum with *Neurospora*. In their work, they selected auxotrophic mutants and identified their nutritional requirements using an ingenious strategy. They first irradiated thousands of fungal spores to induce random mutations. They then placed the spores on complete medium (medium containing vitamins, amino acids, and other complex organic compounds) and allowed each spore to germinate and grow into a multicellular organism (a mycelium). A fragment of each mycelium was then tested for the presence of an auxotrophic mutation by placing it onto minimal medium (medium lacking vitamins, amino acids, etc.). Most fragments did not represent a new, auxotrophic mutant and continued to grow on minimal medium. However, a small number of the fragments failed to grow, and subsequent testing revealed that they were, in fact, auxotrophic mutants.

Beadle and Tatum identified which substance each mutant strain required by further testing additional fragments from the original isolates that continued to grow on complete medium. They tested each mutant strain using a series of steps. For example, first they determined whether a particular strain was deficient in one of a general category of compounds, such as vitamins or amino acids. They did this by attempting to grow fragments from each strain on different kinds of supplemented minimal medium containing groups of compounds. For example, they would use one medium containing a combination of different amino acids and another containing a variety of vitamins.

Q4 If a particular strain tested by Beadle and Tatum turned out to be auxotrophic for one of the amino acids, which of the two types of media described above would support the growth of this strain?

For the unknown auxotrophic strain(s) you will test, assume that they were recently isolated from a mutagenesis experiment and that the preliminary nutritional analyses have already been completed. At this point, your mutant strain(s) has been determined to be auxotrophic either for one of several amino acids or for one of several vitamins (your instructor will tell you which ones). Your task will be to prepare several culture vessels of medium, each supplemented with an individual amino acid or an individual vitamin, and then test your strain(s) on each (and also on complete medium and unsupplemented minimal medium). For each test substance, you should have one culture vessel for each unknown mutant and one for your wild-type control. You should also prepare a culture vessel containing complete medium for each strain (to be used for a control).

Note: As an alternative to using separate culture vessels for each unknown, your instructor may have you grow two or more samples on a single sub-divided Petri plate. Also, your instructor may have you prepare the different types of media as described in Step 2 (below), or, alternatively, you may be given plates of culture media that already contain the components indicated here.

Step 1: Determine how many culture vessels you will need. Also determine which type of medium each will contain (complete medium, minimal medium, minimal medium plus an amino acid, or minimal medium plus a vitamin) and which strains will be grown in each vessel. Label each of the culture vessels appropriately.

Step 2: Prepare minimal medium supplemented with various test substances as needed. Use as much care as possible to maintain sterile conditions during these preparations. If a laminar flow hood (a "sterile hood") is available, prepare and pour your media in the hood. Containers with minimal medium prepared to $0.9X$ volume will be available for your use. It may be necessary to melt this medium using a microwave or hot-water bath. To prepare each type of medium supplemented with a test substance, add a $0.1X$ volume of a $10X$ concentrated stock solution to the $0.9X$ volume of minimal medium. Mix the combined solution and pour some of it into several Petri plates as needed. Allow the supplemented medium to solidify.

Step 3: Prepare culture vessels containing complete medium and unsupplemented minimal medium as needed. Containers with complete and minimal media will be available for your use in preparing your culture vessels. After pouring the medium into one or more Petri plates, allow the complete medium and minimal medium to solidify.

Step 4: After the medium in each of your culture vessels has solidified, transfer cells from each strain to the appropriate culture vessels using a sterile toothpick or probe. Spread the cells somewhat so that growth can be detected after the vessels have incubated for several days.

Step 5: Place your cultures in a 30°C incubator, or leave them at room temperature.

Step 6: Observe your cultures over the next several days. If the cultures are incubated at 30°C, they should exhibit good growth after 1 to 2 days. If they are incubated at room temperature, it may take about twice as long for good growth to occur.

Note: As was the case for the temperature-sensitive strains (see the earlier note), a small number of cells may have reverted to the wild-type phenotype in some of the cultures that originally contained all auxotrophic mutant cells. Therefore, the presence of a small number of growing yeast colonies should not be taken as evidence that the original strain was not an auxotrophic strain. Growth of experimental cultures should always be compared to that of control cultures.

Q5 Which test substance does the yeast in each of your coded cultures require?

Q6 Did any of your auxotrophic mutant strains exhibit any other interesting phenotypes (besides their specific nutrient requirement)? If so, describe the phenotype you observed and speculate on any possible link between this characteristic and the nutrient requirement.

NOTES

NOTES

LAB ACTIVITY

Isolation of
Plasmid DNA

10

INTRODUCTION

This lab activity is unlike all the others in this manual. Each of the other activities posed some problem for you to solve and provided techniques for you to use to solve it. This activity is limited to the techniques only. However, the main technique presented, the plasmid miniprep, is a very useful and common technique in modern genetics. It is presented here because your instructor may choose to have you use this technique to isolate plasmid DNA that will be used in Lab Activity 11, entitled "Genetic Transformation of Yeast" (interestingly, yeast, which is a eukaryote, can be transformed using plasmid DNA from a bacterium!). In this activity, you will isolate plasmid DNA from a liquid culture of bacteria and characterize it via electrophoresis.

PROCEDURE

Isolating the DNA

Step 1: Your instructor will provide you with a sample of an "overnight" bacteria culture containing a dense population of bacteria. Add 1.5 ml of this solution to a sterile microcentrifuge tube.

Step 2: Place your tube in a microcentrifuge so that it is balanced with another tube, and run the microcentrifuge for 1 minute.

Step 3: Remove the supernatant using a micropipette with a sterile pipette tip. A creamy pellet of bacteria should be visible at the bottom of the tube.

Step 4: Add 100 µl of a lysis solution to the tube with the pellet, and resuspend the pellet with a vortex mixer or by repeatedly flicking the tube with your finger. The lysis solution has osmotic properties that help to break the bacterial cells open.

Step 5: Add 200 µl of a solution containing the detergent agent sodium dodecyl sulfate (SDS). This substance is a detergent-like material that will help to solubilize proteins that may be associated with the DNA. Mix the contents by inverting the tube back and forth. During this stage, the cloudy solution should become somewhat clear.

Step 6: Let the solution stand for about 3 minutes, and then add 150 µl of a sodium acetate solution. Sodium acetate will cause the proteins and other substances to precipitate out of the solution. At this point, you will begin to see a white, clumpy material start to form in the solution.

Step 7: Incubate the tube in an ice bath for about 20 minutes.

Step 8: Centrifuge the tube for 5 minutes in a microfuge.

Step 9: Remove the supernatant (approximately 400 µl should be present) using a 200-µl micropipette. Transfer this supernatant into a fresh microcentrifuge tube. This solution contains the plasmid DNA that you are attempting to isolate.

Step 10: Add 1 ml of 95% ethanol to the solution in the clean tube and mix well. The ethanol will cause the DNA to precipitate out of the solution.

Step 11: Centrifuge this solution in microfuge for about 15 minutes. The goal of this step is to obtain a tiny (in some cases, almost invisible) DNA pellet at the bottom of the tube. To ensure that you will be able to locate the DNA pellet after centrifuging, place the tube into the centrifuge in the following way. Place the tube into the rotor with the closure of the tube (i.e., the "hinge") facing downward (i.e., toward the center of the rotor). This way, when the outward force of the centrifuge causes the pellet to form on the lower wall of the tube, you will know where to look for it. The pellet (whether you are able to see it or not) should be located at the bottom of the tube on the opposite side from the closure.

Step 12: Using a micropipette, carefully remove the supernatant and discard it. Place the tip of the micropipette at the bottom of the tube on the same side as the closure when removing the supernatant. This way, you will be less likely to dislodge the DNA pellet.

Step 13: After removing the supernatant, add 500 µl of an 80% ethanol solution. This solution is added as a "wash" solution to help remove materials other than the DNA.

Step 14: Centrifuge the tube as before for about 5 minutes, and carefully remove the wash solution without disturbing the pellet.

Step 15: Now do a final wash with 95% ethanol. Add 500 µl of 95% ethanol. Centrifuge for about 5 minutes and remove the supernatant as before.

Step 16: Remove any remaining ethanol solution by floating the open microcentrifuge tube in a water bath at a temperature between 50° and 65°C for about 10 minutes.

Step 17: Add 100 µl of TE buffer to the pellet to dissolve the DNA. To ensure that the DNA becomes completely dissolved in the TE buffer, repeatedly aspirate the solution by slowly sucking it up and down through a micropipette tip. Do this gently to avoid shearing the DNA.

Step 18: If an appropriately sensitive spectrophotometer is available, your instructor will show you how to use it to determine the concentration and relative purity of your DNA.

First, you will dilute some of your sample into distilled water and use that to obtain your measurement. If possible, you should measure the absorbance of your dilute sample at both 260 and 280 nm (nanometers). You can use the absorbance at 260 nm to calculate the concentration of DNA in your sample, since it is known that 1 absorbance unit equals 50 µg/ml of DNA (remember to factor in the dilution factor when determining the concentration of your original sample). You can also determine the relative purity of your sample by calculating a value for it based on the absorbance value at 260 nm divided by the value at 280 nm. For completely pure DNA, this value will equal 1.8.

Visualizing the Plasmid DNA on an Agarose Gel

Determining the concentration of DNA using a spectrophotometer is one way to see how well your plasmid miniprep worked.

Another way is to run your DNA sample on an agarose gel. If you successfully isolated only the plasmid, and it is mostly intact, you will obtain a single band on your agarose gel (corresponding in size to the approximate size of the plasmid). If your sample is contaminated with other DNA from the bacteria, or if the plasmid was fragmented during isolation, you will see more smearing on your gel. You should run your samples on a gel, as described later. You will then be instructed how to estimate the size of the plasmid, based on the location of the band in the gel.

Step 1: Set aside 20 μl of your plasmid DNA in a microfuge tube to use as a 1X concentrated sample for electrophoresis. Also prepare 20 μl or more of a 0.1X concentration for electrophoresis. Prepare both dilutions for electrophoresis as described here. If your plasmid miniprep DNA is highly concentrated, the bands you see on your gel may be more distinct and reliable for the 0.1X diluted sample.

Your instructor will tell you what volume of DNA solution you need to load the sample wells on your agarose gel. In a separate microfuge tube, mix a small volume of your sample together with a 6X loading dye in the appropriate ratio (i.e., five parts sample to one part 6X loading dye). The dye will give your sample a blue color so that you can see it when you load it into the gel, and it will also make your sample more dense, so that it will sink through the aqueous buffer in the gel tank.

Step 2: Set your micropipette for the volume indicated by your instructor, and remove that amount of your DNA/dye solution.

Step 3: Add your sample to one well in the gel (note which well[s] you load), by submerging the tip below the buffer solution and directly into the opening of a well. Do not expel the solution until you are sure that your tip is inserted into a well, and do not insert your tip too far into the well (be careful to avoid puncturing the bottom of the gel with your pipette tip). When you expel the sample into the well, you should be able to see the colored solution settling into the well and filling it.

One student should also load one lane (on each gel used) with a molecular marker for size determination.

Step 4: After the gel is loaded, your instructor will set the voltage. The gel should be allowed to run until the dye front is close to the opposite end of the gel (unless a somewhat long gel is used).

Step 5: View the gel(s) on an ultraviolet light box or use a gel documentation system to photograph the gel (wear gloves when handling the gel — it contains a potent mutagen, ethidium bromide). If you view the gel directly (rather than viewing an image of the gel), *wear appropriate plastic eye protection*. Since prolonged exposure to UV light may also harm the skin, students may wish to wear a plastic face shield and appropriate clothing to cover exposed skin.

Measure how far each band migrated from the well. You will use this information to determine the size of the plasmid (see next step).

Measure the distance from the well to the location on the gel where your plasmid seems to be located (if several bands or a smear are present, measure the largest and the most distinct band[s] present; measure the distance to the center of the bands). Also measure the distances between the well and each of the bands in the lane containing the reference DNA. Your instructor will tell you the sizes of the DNA fragments represented by each band in the reference DNA lane.

Step 6: To estimate the size of the DNA you isolated, compare the distance traveled by the band representing the plasmid to that of each of the reference bands. Do this by plotting a standard curve for the reference DNA fragments on semilog graph paper and comparing the measurement for your plasmid DNA to these. Plot the sizes of the reference fragments (in base pairs) on the log scale axis and the distances migrated on the standard scale axis. You should be able to draw a straight line passing through (or nearby) each of the points. Now use the standard curve to estimate the size of the plasmid you isolated.*

Q1 **How large was the plasmid you isolated according to the estimate from the standard curve? Hand in a copy of your standard curve showing distance migrated by your plasmid and its estimated size on the graph paper on the next page.**

* Note: If you isolated the plasmid without breaking the circular DNA molecule, it will be closer to the bottom of the gel than you would predict, based on its known size. This occurs because uncut plasmids are supercoiled, and supercoiled DNA travels through an electrophoresis gel faster than linear DNA of the same size.

LAB ACTIVITY 11

Genetic Transformation
of Yeast

INTRODUCTION

Transformation is the term used to describe the process whereby a cell takes up foreign DNA that may then be incorporated into its genome and even be expressed. Transformation was first accomplished with bacteria in the early 1900s, but has since been extended to yeast, animal cells, and, with some modifications, even plant cells. In this lab activity, you and others in your lab will attempt to transform several auxotrophic yeast strains (like those studied in a previous lab) using a plasmid that carries a yeast gene (the plasmid was isolated from a bacterial host using the procedure described in a separate lab activity). You will also investigate some factors influencing the efficiency of the transformation process.

The yeast strains you will transform are all auxotrophic for production of the nitrogenous base adenine (which is also considered to be a vitamin, called vitamin B_4). The transformation experiment you will be doing is somewhat like conducting "gene therapy" for yeast cells — you will be trying to provide each mutant strain with a wild-type gene for adenine biosynthesis (enabling it to survive on minimal medium, without added adenine). The wild-type form of the adenine biosynthesis gene is carried on a genetically engineered, bacterial plasmid. Yeast are especially good at taking up and expressing foreign DNA. Even though the plasmid that carries the yeast gene was originally taken from a bacterium, it will be easily taken up and expressed by the yeast used in this activity. However, the efficiency of the transformation process (i.e., the number of cells transformed) will be quite different for different groups of student researchers.

The efficiency of yeast transformation varies significantly from strain to strain, and even somewhat from experiment to experiment when the same strain is used. One possible reason for this variation could be differences in growth rates (between strains and within strains in different experiments). If one strain grows faster than another, or if experimental conditions change, resulting in different growth rates, the number of cells and their physiological condition will also be different. These factors almost certainly affect transformation efficiency. On the other hand, differences besides those related to growth rate may also affect transformation efficiency. For example, different strains may have different kinds of cell surface receptors that may affect their ability to take up foreign DNA.

In this lab activity, different groups of students will attempt to transform different strains of adenine auxotrophs, and each group will attempt the transformation using different volumes of cell suspensions. You will then determine the relative transformation efficiency for each strain, and for each of the volumes used by counting the numbers of colonies produced in each case. To count the numbers of colonies produced, you will need to dilute the various cell samples (i.e., pretransformation cells and each sample of transformed cells) into several concentrations. When the experiment is complete, the entire class will share the data, and you will attempt to determine the relative importance of the differences between strains, as compared to differences in cell numbers.

PROCEDURE

You will conduct the following transformation procedures and prepare cultures to assess transformation efficiency in small student research groups. Different groups will work with different adenine auxotrophic mutants, and each group will attempt three transformations with different volumes of yeast cells.

Step 1: Preparing dilutions to determine the initial concentration of yeast cells — Your instructor will provide you with a sample of an "overnight" yeast culture grown in a solution of complete medium. This solution contains a dense population of auxotrophic yeast cells that will serve as the source of the cells that you will transform. To determine the transformation efficiency for your experiments, you will first need to know the initial concentration of the cells in this overnight culture. You will determine this number by growing three cultures on complete medium at various densities and counting the colonies that result from one of the dilutions (by using three dilutions, you should obtain one that produces an optimum number for counting).

Before preparing the cultures, you will need to dilute your yeast sample so that the concentration of cells will be small enough to allow resulting colonies to be counted. Prepare 1 ml of a 100-fold dilution and 1 ml of a 1,000-fold dilution as follows (using sterile technique as much as possible). Prepare the dilutions in labeled, sterile microcentrifuge tubes using sterile, distilled water.

Make the 100-fold dilution by adding 10 µl of yeast suspension to 990 µl of sterile, distilled water. Make the 1,000-fold dilution by adding 100 µl of the 100-fold dilution to 900 µl of sterile, distilled water. Be sure to fully suspend each of the solutions of cells before taking aliquots for dilution to ensure that a representative density of cells is obtained.

After preparing the two dilutions, you will be ready to prepare your three cultures. First label three culture plates containing complete medium as follows. The labels should include notations to indicate that the medium is complete medium, that the cells are from a specific strain, that they are "nontransformed" cells, and that one of three dilutions, a, b, or c, will be added to the culture medium.

Then add 2 to 3 drops of sterile, distilled water to the surface of the medium in each plate to facilitate spreading the yeast suspension. Next, add the following volumes of your cell dilutions directly into the sterile drops of water: (a) 1 µl of the 1,000-fold dilution, (b) 10 µl of the 1,000-fold dilution, and (c) 10 µl of the 100-fold dilution. Then use a sterile spreading loop to thoroughly spread the yeast cells over the entire surface of each plate.

Also prepare three additional culture plates containing minimal medium as described earlier, using the same concentrations of yeast suspensions, and label the plates appropriately. These will serve as control cultures for comparison to the cultures you will make after transforming the yeast (Step 8).

Place all six cultures described so far (along with the ones you will prepare in Step 8) into a 30°C incubator, if available, or incubate them at room temperature.

Step 2: Remove three aliquots of the overnight yeast culture to be transformed, and place them into labeled, sterile microcentrifuge tubes. Each aliquot will contain a different volume of cells (and, hence, a different number of cells). Be sure to fully suspend the solution of cells before taking each aliquot to ensure uniform cell density. These separate aliquots will be treated identically in the subsequent transformation steps, and the results will be used to evaluate the effect of varying the number of cells.

One aliquot should contain 500 µl of cells, a second aliquot should contain 1000 µl, and the third aliquot should contain 1500 µl.

Step 3: Place these tubes into a microcentrifuge with the tubes balanced against tubes from other student groups (balance tubes with the same volumes, and ensure that tubes are labeled so that they can be identified). Run the microcentrifuge for 20 seconds. You will see a dense pellet of yeast cells at the bottom of each tube.

Conduct each of the following steps for all three of your samples in an identical manner.

Step 4: Carefully remove the supernatant leaving an undisturbed pellet of cells at the bottom of the tube. You should use a micropipette with a sterile pipette tip to remove the last portion of the remaining solution. If you notice that some cells are being withdrawn during this process, it is better to leave behind a small amount of solution, rather than disturb the pellet.

Step 5: Add 100 µl of a solution that contains the following: 0.2 M lithium acetate, 40% polyethylene glycol-4000, and 100 mM DTT. Resuspend the pellet with a vortex mixer or by repeatedly aspirating the solution with the micropipette.

The lithium acetate affects cell permeability (to enhance DNA uptake), and the polyethylene glycol promotes interactions between the DNA (added in the next step) and the cell surface (the DTT is a chemical stabilizer).

Step 6: Add 10 µl of plasmid DNA and 5 µl of DNA that is called "carrier" DNA to this solution. Add these volumes very carefully, since changes in volume of the DNA will affect transformation efficiency. Be especially careful when pipetting the carrier DNA since it may be quite viscous.

Single-stranded salmon sperm DNA is often used as carrier DNA. The genes on this DNA will not be expressed by the yeast cells, but the presence of this "extra" DNA somehow facilitates the uptake of the plasmid DNA.

Step 7: Vortex this mixture for several seconds (or mix it thoroughly by repeated aspiration with a micropipette) and then place the tube in a floating rack in a 45°C water bath for 30 minutes. This is the final step in the transformation process. Heating the cells and the DNA apparently creates a dynamic condition that promotes DNA uptake.

Step 8: After the 30-minute incubation period, you will prepare two dilute samples for each tube of transformed yeast and spread aliquots from both dilutions onto one or more Petri plates containing minimal medium (without adenine). It is necessary to prepare dilutions so that the cell density will be low enough that single cells can form discrete colonies. You will prepare two dilutions of the transformed yeast because some samples may produce too many transformed cells to count at the higher dilution, while other samples may produce too few to count accurately at the lower dilution. Prepare the dilutions and spread them on Petri plates with minimal medium, according to the following instructions.

Each tube should contain 115 µl of solution (100 µl of transformation solution with yeast cells, 10 µl of plasmid DNA, and 5 µl of carrier DNA). Add enough sterile, distilled water to each sample to bring the volume up to the original aliquot of cells (i.e., add 385 µl to bring the first sample up to 500 µl, add 885 µl to bring the second sample up to 1,000 µl, and add 1,385 µl to the third sample to bring it up to 1,500 ul). Prepare a second dilution for each sample that is 10 times less concentrated by mixing 100 μl of the first dilution (make sure it's well suspended) with 900 μl of sterile, distilled water.

Now prepare one culture for each dilution of your three samples of transformed cells on minimal medium, as follows. For each culture, add 100 μl of the cell dilution to the surface of the medium, and spread the solution thoroughly over the entire surface. (*Note:* If you wish to obtain a more accurate estimate of the number of transformed cells, you can prepare multiple cultures in the same fashion and count the resulting colonies from each plate.)

Step 9: Place these cultures (along with the ones prepared in Step 1) in a 30°C incubator for 3 to 4 days, or allow them to grow at room temperature for somewhat longer.

Step 10: The following describes the rationale for determining the transformation efficiency for each sample of your yeast strain. Obviously, this analysis must be performed after the incubation step described in Step 9. To determine transformation efficiency, you will need to determine two values: (1) the number of cells per milliliter in the original culture (pretransformation) and (2) the number of transformed cells in each sample. You will only need to determine these values for one dilution in each case. Some dilutions will produce too many colonies to count accurately, and others will produce too few. If possible, select a dilution that yields more than a few colonies per plate, but not one in which the colonies overlap one another. To count the colonies, you should draw a grid on the bottom of the culture plate (using a permanent marker or a wax pencil) and count the colonies in each square of the grid to avoid overlooking any.

The following describes each type of culture that you will generate in this experiment and gives you instructions for collecting and evaluating data from each. In all, you will have at least 12 different cultures (more, if you prepared multiple plates from your transformed cells).

Three cultures will represent pre-transformation cells at three different densities (a, b, and c) on complete medium. You should count the colonies produced by cells at one of these dilutions, and determine *the average number of cells per milliliter* in the original overnight culture (remember to take into account the necessary dilution factors).

Three cultures will represent pre-transformation cells at three different densities (a, b, and c) on minimal medium. You should count the colonies produced (if any) by cells at one of these dilutions, and determine *the average number of cells per milliliter* from the original overnight culture that were able to survive without adenine (remember to take into account the necessary dilution factors).

At least six cultures will be prepared from transformed cells (more, if you prepared multiple plates from these cells). For each volume of cells (500 μl, 1000 μl, and 1500 μl), you will have two cultures prepared from different dilutions. You should count the colonies produced by cells at one of these dilutions, and estimate *the number of cells that were transformed* for each 1 ml (ie., 1000 μl) of the original aliquot. Remember that you only tested a portion of each sample, and you will have to estimate the total number of transformed cells based on the amount you tested. Also, remember to account for the dilution factor if you counted colonies from the diluted sample.

Q1 What was the average number of cells per milliliter in your overnight sample? Show your calculations, indicating with notes what each number in your calculation represents.

Q2 A. Did any colonies appear on the minimal medium containing nontransformed cells? If so, how many such cells per milliliter were present in the original overnight culture?

B. It is reasonable to expect that you may have observed a very small number of nontransformed cells growing on minimal medium. Propose a reasonable hypothesis explaining why this kind of result might be expected to occur from time to time.

Q3 For the transformed cells, compare the number of cells per ml that survived on minimal medium for the various strains tested by different groups in your class. If any colonies grew on minimal medium for a given strain, subtract the number of colonies per ml that grew on minimal medium from the apparent number of transformed cells per ml. Also compare the number of transformed cells per ml for the different volumes (i.e., cell numbers) for these strains. Prepare a table that shows this information. You will be instructed to evaluate these data in following questions.

Q4 Is there an apparent difference in the transformation efficiency for the various adenine auxotrophic strains tested by your class? Describe the rationale for your answer to this question, citing specific examples.

Q5 Is there an apparent difference in the transformation efficiency when different numbers of cells are used (i.e., the three different volumes should have proportionally different numbers of cells)? Respond to the following considerations as you answer this question.

One might expect to see some increase in the number of transformed cells as the initial number of cells used for the transformation increases. Did this trend occur for most (all?, some?, any?) strains tested by your class, or was a different general trend observed? If an increase occurred, was there generally a linear relationship between increases in the initial number of cells and increases in the number of transformed cells? Illustrate the relationship (linear or otherwise) between these factors for several strains by graphing the data (volumes of cell suspensions vs. numbers of cells transformed). Describe what factors (cell density, DNA concentration, etc.) may contribute to the trend you observed and discuss specifically how these factors may be related to the trend (i.e., formulate a hypothesis).

Q6 If differences in transformation efficiency existed between the different strains studied by your class, do the differences seem to be related primarily to differences in growth rates (as indicated by different cell densities in overnight cultures) or to some other factor(s)? Explain your answer with reference to specific data from the table you prepared in response to Question 3.

LAB ACTIVITY

Isolation of Novel Genes
from a cDNA Library

12

INTRODUCTION

In this laboratory activity, you will utilize techniques used by genetic researchers to discover previously unidentified genes from eukaryotic organisms. The lab may be conducted as a demonstration (i.e., screening for known genes) or as a novel experiment (screening for unknown genes). The kind of library you will work with (i.e., human, *Arabidopsis*, *Ceratopteris*, etc.) and the gene you attempt to identify will be determined by your instructor. If your class is able to conduct a screen for previously unknown genes, this experiment could lead to additional studies (a student and/or faculty research project) and the results could ultimately be published! Your goal for now will be to use an antibody-based selection scheme to identify bacterial hosts infected with viral vectors carrying a specific gene.

The library you will use was constructed by inserting cDNA copies of mRNAs from a eukaryotic organism into bacteriophage vectors (also simply called *phage vectors*). *Bacteriophages* are viruses that specifically infect bacteria, and *vectors* are entities that carry and replicate foreign DNA. The phage vectors in this library have been engineered in such a way that the foreign genes will be expressed (i.e., both transcribed and translated). To identify and isolate particular genes in this library, you will use a selection scheme that employs antibodies against these proteins (described in detail later). The cDNA library contains copies of many genes (besides the ones you are trying to identify), but because of the specificity of the detection assay (antibody-based), only bacteria with the virus containing the genes of interest will be detected.

It will require several days to conduct all aspects of this experiment (described later under "Procedures"). The class will divide the responsibility for conducting different parts between various groups of students. Your instructor may choose to have some steps of the experiment done by a lab assistant, rather than assigning these responsibilities to student groups. You should read the following description of the selection scheme and familiarize yourself with all of the steps before coming to lab. Your instructor will then explain how your class will conduct the specific steps of the experiment.

Note: A good deal of reading should be done prior to coming to lab. The amount of reading is more than usual, but it is *essential* to your understanding that you read the following to prepare for lab. Otherwise, this will seem like a "cookbook" exercise.

THE SELECTION SCHEME

Eukaryotic organisms possess tens of thousands of genes, and a given cDNA library may contain copies of most or at least a large percentage of these genes. Once a cDNA library has been constructed for a given organism, the next challenge is finding genes of interest. There are several ways of doing this, but all of them involve growing many, many random representative vectors from the library and somehow screening them all to isolate a small number that carries the gene of interest. A screening strategy often described in introductory textbooks involves finding the gene of interest by "probing" the library with a labeled polynucleotide. This approach has been used quite successfully, but it has certain limitations, including the fact that some sequence information for the gene of interest must be available (to design the polynucleotide probe). You will be using a different approach that employs antibodies against the protein made from the gene of interest as the probe. This eliminates the requirement for specific sequence information prior to screening.

Q1 **Assume that you want to screen a cDNA library for a gene that has not yet been sequenced. How would the antibody for the protein from this gene be generated?** (*Note:* **You could provide a highly detailed answer to this question; however, an answer of one to several sentences will be sufficient.**)

The multistep procedure for antibody-based screening (see Figure 12.1) is described in detail later. Briefly, it involves the following elements: (1) growing the phage-infected bacteria in a layer of "top agar" (described below) on nutrient agar plates, (2) stimulating gene expression with isopropyl ß-D-thiogalactoside (IPTG)-soaked filters and allowing expressed proteins to adhere to the filters, (3) incubating the filters with the primary antibody (which recognizes the protein of interest), (4) incubating the filters with a secondary antibody (which recognizes the primary antibody and carries an enzyme "label"), (5) "developing" the filters (the enzyme linked to the secondary antibody causes a color change), and (6) identifying the location of the gene of interest on the culture plates and removing the bacteriophages from the culture at that location.

Note that this detection method can only be used successfully for genes that are present in eukaryotes but not in bacteria. Otherwise, the antibody would cross-react with bacterial gene products.

Growing Phage-Infected Bacteria

In this part of the protocol (Steps 1 to 7 in the Procedures), bacteria and phage vectors are combined in predetermined quantities such that hundreds of thousands of infections can take place. This is necessary to increase the likelihood of finding the gene of interest. The phage-infected bacteria are grown in such a way that the culture plates are covered with a "lawn" of bacteria (growing in an upper layer known as *top agar*) with infected cells scattered throughout the lawn. As the cultures are incubated, the infections spread, forming clones of infected bacteria that are visualized as "plaques" on the bacterial lawn. A plaque is a clear spot in the top agar where all of the bacteria have been lysed. Your instructor will determine the appropriate concentration of bacteria and phages for your experiment. It is ideal to have so many plaques on a culture plate that they are densely packed together, yet few enough that it is still possible to identify individual plaques.

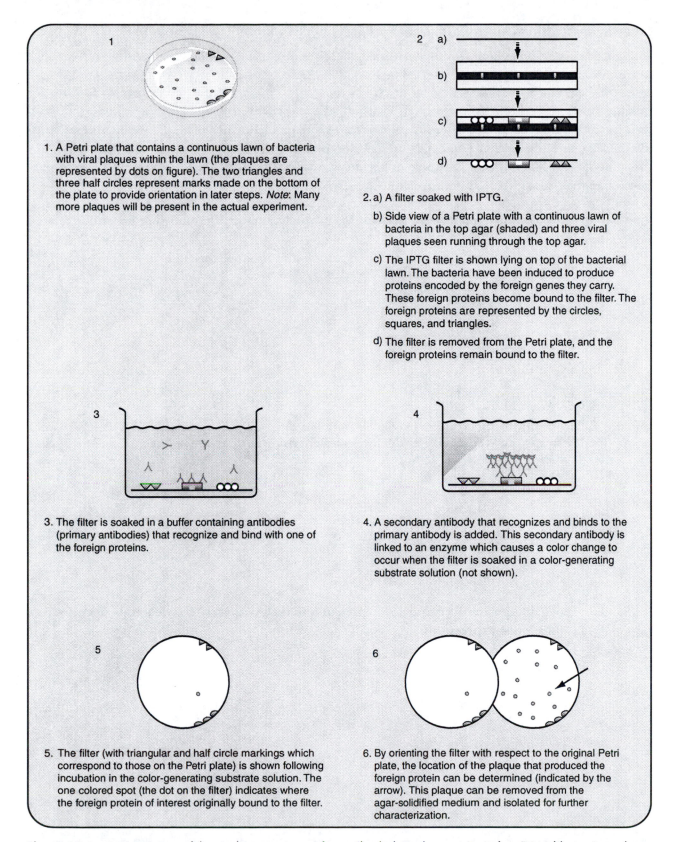

1. A Petri plate that contains a continuous lawn of bacteria with viral plaques within the lawn (the plaques are represented by dots on figure). The two triangles and three half circles represent marks made on the bottom of the plate to provide orientation in later steps. *Note*: Many more plaques will be present in the actual experiment.

2. a) A filter soaked with IPTG.

   b) Side view of a Petri plate with a continuous lawn of bacteria in the top agar (shaded) and three viral plaques seen running through the top agar.

   c) The IPTG filter is shown lying on top of the bacterial lawn. The bacteria have been induced to produce proteins encoded by the foreign genes they carry. These foreign proteins become bound to the filter. The foreign proteins are represented by the circles, squares, and triangles.

   d) The filter is removed from the Petri plate, and the foreign proteins remain bound to the filter.

3. The filter is soaked in a buffer containing antibodies (primary antibodies) that recognize and bind with one of the foreign proteins.

4. A secondary antibody that recognizes and binds to the primary antibody is added. This secondary antibody is linked to an enzyme which causes a color change to occur when the filter is soaked in a color-generating substrate solution (not shown).

5. The filter (with triangular and half circle markings which correspond to those on the Petri plate) is shown following incubation in the color-generating substrate solution. The one colored spot (the dot on the filter) indicates where the foreign protein of interest originally bound to the filter.

6. By orienting the filter with respect to the original Petri plate, the location of the plaque that produced the foreign protein can be determined (indicated by the arrow). This plaque can be removed from the agar-solidified medium and isolated for further characterization.

Figure 12.1 – An overview of the multistep process for antibody-based screening of a cDNA library in a phage expression vector. See the text and figure notes for details.

Overlaying Cultures with IPTG-Soaked Filters

The phage expression vectors you will be using can be stimulated with the chemical IPTG to induce expression of their foreign genes (Steps 8 to 11 in the Procedures). They have been engineered so that a genetic element that functions as an "on/off" switch has been placed into the vector DNA just prior to where the foreign DNA is inserted. This on/off switch is the well-known *lac operon* that some bacteria use to naturally regulate the production of lactose-metabolizing enzymes. If you have never studied the lac operon before, or you have, but have forgotten what you learned about it, you should take time now to review this classic example of genetic regulation. Most up-to-date general biology textbooks describe this system, as do virtually all genetics textbooks.

 Now that you have reviewed the lac operon, briefly answer the following questions related to its function:

A. What is the "inducer" in the lac operon system?

B. What is the "repressor" in the lac operon system?

C. How do the inducer and the repressor interact with a site called the "operator" to induce gene expression?

The chemical IPTG is a synthetic compound that takes the place of lactose for the purpose of this experiment. Unlike lactose, it is not metabolized by the structural genes of the lac operon; however, it is still able to "turn on" gene expression. After the phage-infected bacteria have been growing for several hours, filters soaked with IPTG are laid on top of the culture medium. The filters and culture plates are marked so that after the filters are removed, they can later be realigned with the culture plate (described later). The IPTG-soaked filters are made of a material that has a high affinity for proteins. Thus, when the IPTG stimulates the expression of the foreign genes, the gene products (proteins) become adhered to the filter. At this point, it becomes possible to manipulate the filters to screen for evidence of the gene of interest, while allowing the bacteria that are infected with phage vectors to remain on the culture plates. After the location of the gene of interest is identified on the filter, the filter can be realigned with the culture plate to determine where the "positive" plaque is located.

The steps described so far can be accomplished in one day. The remainder of the experiment can be accomplished on the following day or over a period of days thereafter.

Incubating Filters with the Primary Antibody

This part of the experiment (Steps 12 to 19 in the procedures) involves washing and blocking the filters (discussed later) and allowing them to react with the primary antibody. It is the reaction with the primary antibody that provides the specificity in this experiment so that only certain genes are identified by the screen. Any of a very large number of possible antibodies could be used during this step; however, you should follow two main guidelines to select an antibody that is most likely to provide positive results. First, the antibody should be one that reacts only with proteins produced by eukaryotic organisms. Otherwise, if bacteria also produce this protein, all of the bacteria growing on plates in this experiment will produce a positive signal. Second, the antibody should have been made from a protein that should be similar to a protein in the organism from which your library was made.

There are several ways that this criterion can be met. For example, the antibody could be specific for a protein (or a region of a protein) that appears to be highly conserved. In such cases, it is possible that an antibody made for an organism in one kingdom (e.g., humans), could even react with proteins from an organism in another kingdom (e.g., a fern). Or, the antibody could be specific for a protein from the actual species for which your library was made or for a closely related species.

Another consideration in selecting the antibody is related to the likely abundance of the mRNA for that protein in your organism. Since the cDNA library was made from mRNA, the number of copies of a given type of cDNA present in your library will be based on the original level of that mRNA in your organism. For example, the number of copies of tubulin cDNAs (a high-abundance "housekeeping" gene) in a given library is likely to be much higher than the number of tyrosine kinase cDNAs (tyrosine kinases are regulatory genes). This means that you are more likely to have success if you choose to search for a gene that is highly transcribed. However, on the other hand, the less commonly transcribed genes are often much more interesting!

It is also possible to conduct this experiment with built-in controls, so that obtaining some positive plaques is almost guaranteed. For example, you could screen one library for a previously unidentified gene, but include some isolated phages (on a control culture) from another library that are known to carry that gene. If such a control sample of phages is available, this step is recommended.

Many other steps conducted before and after addition of the primary antibody are either "washing" or "blocking" steps. Both of these types of procedures help to reduce the amount of background signal (i.e., they help eliminate false positives). The washing steps are necessary to remove material that has become loosely adhered to the filter in a nonspecific way. The blocking step uses milk proteins to "cover" unbound areas of the filter (places where no foreign gene has adhered). This step is done just prior to adding the primary antibody so that the antibodies will not adhere to the empty spaces on the filter in a nonspecific way.

A step in this part of the experiment also involves gluteraldehyde fixation. During this step, the gluteraldehyde interacts with the protein and the antibody to create covalent bonds between them. This ensures that the antibody will remain bound to the filter during the remaining steps.

Incubating Filters with the Secondary Antibody

Steps 20 to 22 in the Procedures

The secondary antibody is "tagged" with an enzyme (a peroxidase) that causes a color reaction to occur (see the next section) on the filter where protein from positive plaques is located. The secondary antibody (usually an antibody made in a goat) recognizes the primary antibody (usually an antibody made in a mouse) and binds specifically to it. Without adding the secondary antibody, the primary antibody cannot be detected (no color reaction will occur).

There are also washing and blocking steps associated with the incubation with the secondary antibody. These steps serve essentially the same purpose as the previous washing and blocking steps associated with the primary antibody.

Developing the Filters

Steps 23 to 25 in the Procedures

During this stage of the experiment, the filters are exposed to a solution that contains a colorless substrate for the peroxidase enzyme on the secondary antibody. As the enzyme interacts with this substrate, it transforms it into a bluish/purple-colored product. This product becomes localized on the filter.

Locating the Gene of Interest and Isolating Positive Plaques

Steps 26 to 29 in the Procedures

After the location of the foreign protein for the gene of interest has been identified on the filter, it is necessary to use the filter to locate the positive plaque on the experimental culture plate(s). This is accomplished by using filters and culture plates marked with corresponding unique indicators (i.e., nicks or holes in the filters and corresponding permanent pen markings on the bottoms of the culture plates). After positive spots have developed on one or more of the filters, plastic sheets (i.e., overhead transparencies) are used to make "maps" of the filters, showing where the positive spots occurred.

These are then placed beneath the culture plates and used as guides for removing "plugs" from the culture medium that contains the positive plaques. The plugs are removed using the wide ends of glass pipettes and then transferred into a phage elution buffer. The plugs will also probably contain other plaques (which lack the correct foreign gene), and further "subcloning" (repeating this procedure with material from these plugs) will be necessary to purify the sample. However, this lab activity will be complete after the plugs containing positive plaques have been removed.

Q3 Now that you have read the description of this experiment, explain in your own words why two antibodies are used in this experiment.

A. What is the specific role of each of the two antibodies?

B. Can you think of any advantages to using two antibodies and having the second one be "enzyme linked," rather than using an enzyme-linked primary antibody? Explain. (*Hint:* As you try to answer part B of this question, it may be helpful to consider the implications of part 4 of Figure 12.1.)

PROCEDURES

It will require several days to conduct all aspects of this experiment as described here. The class will divide the responsibility for conducting parts of the experiment among various groups of students. Your instructor may choose to have some steps done by a lab assistant, rather than assigning these responsibilities to student groups. *Before beginning, be certain that you know which parts of this experiment you are responsible to conduct!*

Specific Steps for Phage Plaque Screening:

Steps 1 to 7: Growing Phage-Infected Bacteria

Step 1: Grow an overnight culture of the bacteria used for screening the library. On the following day, read the OD_{600} of this culture in a spectrophotometer to estimate the initial density.

Step 2: Centrifuge the overnight culture 10 minutes at 2,000 rpm.

Step 3: Decant the supernatant, and suspend the pellet in a 10 mM $MgSO_4$ solution to a final OD_{600} of 0.5 (this value may be changed for some strains).

Step 4: Mix aliquots of the phage library to 600 µl of cells in disposable 10- to 15-ml tubes for screens conducted on 150-mm culture plates (200 µl for 100-mm plates). *Note:* Your instructor will determine how many microliters of the phage library to add.

Step 5: Incubate the cells and phages at 37°C for 15 minutes.

Step 6: Add 6.5 ml of warm top agar (3 ml for 100-mm plates) to each tube of infected bacteria, and spread evenly over a plate containing NZY medium that was kept in a 37°C chamber just before use.

Step 7: Incubate the plaques at 42°C until small plaques form (ca. 3.5 hours).

Steps 8 to 11: Overlaying Cultures with IPTG-Soaked Filters

Step 8: Make asymmetrical, unique nicks (e.g., cut triangles and/or circles) on edges of the filters and autoclave them. After removing the filters from the autoclave, attempt to keep them as sterile as possible before placing them onto the cultures (use sterile IPTG, handle with sterile forceps, etc.).

Step 9: Treat the filters with 10 mM IPTG solution at least 30 minutes prior to use, and wet the filters by submerging them in the IPTG solution until they are completely wet. Place them slowly into the solution, starting at the edge, and allow the solution to be drawn into the filters by capillary action. Put the filters onto blotting paper briefly to dry them somewhat (the filters should remain moist, but not be sopping wet).

Step 10: Place filters on the surface of the culture plates, and mark the bottoms of the plates with lines corresponding to the marks on the filters. These corresponding marks should be made as carefully as possible so that any positive plaques can be easily isolated at the end of the experiment.

Step 11: Incubate the plates at 37°C for 3.5 hours. After this, the plates with filters can be placed in a cold room or refrigerator overnight.

Steps 12 to 19: Incubating Filters with the Primary Antibody

During many of the remaining steps, the filters will be soaked in various solutions. The volume of each solution should be about 10 ml per filter. While the filters are soaking in solutions, they should be rocked gently, using a shaker or rocking platform.

Step 12: Carefully remove filters with forceps, and wash them in TBST solution. Immerse the filters in TBST and remove any remaining top agar with a gloved hand (wash gloves first) or smooth, metal rod. Combine the filters in a plastic container or a heat-sealed bag, and wash 3 to 5 times with TBST for at least 15 minutes each.

Step 13: Immerse the filters in blocking solution (5% weight/volume dry milk in TBST) and agitate gently for 1 hour at room temperature to block remaining protein binding sites. (Filters can be stored for at least several days at this point in blocking solution with 0.02% sodium azide weight/volume at 4°C in a sealed bag. *Sodium azide is poisonous; handle with care!*)

Step 14: Transfer the filters into a primary antibody solution. Incubate with gentle shaking for at least 1 hour at room temperature. Turn (i.e., "flip") and spread the filters periodically.

Step 15: Wash three times in 0.05% Tween 20 (v/v) in PBS for at least 15 minutes each wash. Phosphate buffered saline (PBS) rather than Tris buffered saline (TBS) is used in this part of the procedure because the *Tris* interferes with the reaction mediated by the gluteraldehyde in the next step.

Step 16: Fix the primary antibodies to the proteins by washing with 0.1% gluteraldehyde in cold PBS for 15 minutes. *Handle the gluteraldehyde with care — it is toxic!*

Step 17: Wash once in PBS for 15 minutes.

Step 18: Block with 5% dry milk in PBS for 20 minutes.

Step 19: Wash twice in 0.05% Tween 20 (v/v) in PBS for at least 15 minutes each wash.

Steps 20 to 22: Incubating Filters with the Secondary Antibody

Step 20: Transfer the filters into 10 ml per filter of fresh TBST blocking solution containing peroxidase conjugated antimouse antibody, and incubate with gentle agitation for 1 hour at room temperature.

Step 21: Wash the filters 3 to 5 times in 10 ml per filter TBST for 5 minutes each time to remove any residual unbound or nonspecifically bound enzyme-conjugated antibody.

Stop 22: Remove residual Tween 20 by doing a final wash in 10 ml per filter TBS alone.

Steps 23 to 25: Developing the Filters

Step 23: Prepare 90 mg 4-chloro-l-naphtol in 60 ml methanol, and add this to 300 ml TBS along with 120 μl of H_2O_2 (this volume is sufficient for up to about 20 filters).

Step 24: Filters are immersed in this substrate mixture and incubated at room temperature with gentle shaking. Staining of positive plaques can usually be detected within 15 to 25 minutes. Proteins from different plaques will stain with different intensity (depending on the specific structure of the protein and antibody binding). The vast majority of plaques should not produce a positive reaction. True positive spots will appear uniquely dark compared to most neighboring spots (however, slight background staining is likely).

Step 25: Stop reaction by washing with water.

Steps 26 to 29: Locating the Gene of Interest and Isolating Positive Plaques

Step 26: Dry the filter(s) that indicate the presence of positive plaques, and use a plastic sheet (i.e., an overhead transparency) to draw the outline of the filter (including the unique nicks on the edges) and the location of the positive plaque(s). Do this very carefully so that you will be more likely to actually isolate the positive plaques in the next step.

Step 27: Position the culture plates on the plastic sheets that show the locations of positive spots, so that the marks on the plates and marks on the plastic sheets are aligned.

Step 28: Using the wide end of a sterilized glass pipette, and, maintaining sterility as much as possible, remove a plug of the culture medium that should contain the positive plaque (based on the mark on the plastic sheet beneath the plate).

Step 29: Transfer this plug into 1 ml of sterile SM buffer containing 2 drops of chloroform.

The phages present in this plug of culture medium will diffuse out into the SM buffer after about 24 hours. Although the plug you isolated probably contains several phages carrying different foreign genes, the gene of interest should be present at a relatively high titer. It will now be possible for someone eventually to isolate only the phage carrying the gene of interest by repeating the prior protocol using various dilutions of the phage you isolated. The ultimate goal will be to sequence this newly discovered gene and so contribute to our existing knowledge about the structure and variation of eukaryotic genes.

Q4 For this question, your instructor will ask you to respond for a particular organism of his or her choice. Assume, that you have a cDNA library for the organism your instructor has indicated. If you could do this experiment again using this cDNA library and any antibody of your choice, what gene would you try to identify and why? In your answer, include technical reasons that you would choose a particular gene (how likely is it to work?), and also describe what scientific value might be associated with isolating this gene (i.e., how could your discovery be used in the future?). You should try to identify a gene that has not already been found and sequenced. To help you identify such a gene, go to the National Center for Biotechnology Web page for Nucleotide searches (www.ncbi.nlm.nih.gov/sites/entrez?db=Nucleotide), and verify that the gene you're interested in hasn't already been found.

NOTES

NOTES